AF663845

Discrete Mathematics

Discrete Mathematics

Sudarsan Nanda
Professor of Eminence and Research Chair
KIIT Deemed to be University
Bhubaneswar 751024, Odisha, India
Retired Professor of Mathematics, IIT Kharagpur
Former Vice Chancellor, North Orissa University

ALLIED PUBLISHERS PVT. LTD.
New Delhi • Mumbai • Kolkata • Chennai • Bangalore • Hyderabad

ALLIED PUBLISHERS PRIVATE LIMITED

D-5, Sector-2, **Noida**–201 301
Ph. Nos.: 0120-4320295/2542557/4352866 • E-mail: delhi.books@alliedpublishers.com

17 Chittaranjan Avenue, **Kolkata**–700072
Ph.: 033-22129618 • E-mail: cal.books@alliedpublishers.com

15 J.N. Heredia Marg, Ballard Estate, **Mumbai**–400001
Ph.: 022-42126969 • E-mail: mumbai.books@alliedpublishers.com

No. 25/10, Commander-in-Chief Road, Ethiraj Lane (Next to Post Office)
Egmore, **Chennai**–600008
Ph.: 044-28223938 • E-mail: chennai.books@alliedpublishers.com

P.B. No. 9932, No. 15, 3rd Floor (Next to Vijaya Bank), 5th Cross, Gandhinagar, Karnataka, **Bangalore**–560009, Ph.: 080-41530285 / 22386239
• E-mail: bngl.journals@alliedpublishers.com / apsabng@airtelmail.in

Sri Jayalakshmi Nilayam, No. 3-4-510, 3rd Floor (Above More Super Market)
Barkatpura, **Hyderabad**–500027
Ph.: 040-27551811, 040-27551812 • E-mail: hyd.books@alliedpublishers.com

Website: www.alliedpublishers.com

© 2022, Sudarsan Nanda

ISBN: 978-93-90951-18-5

No part of the material protected by this copyright notice may be reproduced or utilized in any form or by any means, electronic or mechanical including photocopying, recording or by any information storage and retrieval system, without prior written permission from the copyright owners. The views expressed in this volume are of the individual contributors, editor or author and do not represent the view point of the centre.

Published by Sunil Sachdev and printed by Ravi Sachdev at Allied Publishers Pvt. Ltd.,
D-5, Sector-2, Noida–201 301

Preface

This book is suitable as a text on 'Discrete Mathematics' for undergraduate engineering students of computer science, Mathematics (Hons) and for MCA students of all Indian Universities and engineering colleges. This is written at par the model syllabus of UGC which has been accepted by several Universities of India including all the Universities of Odisha state.

There are several books on 'Discrete Mathematics'. However, there are not many titles published by Indian publishers or low-priced titles by foreign publishers which are as per the syllabus of Indian Universities. Some of them are elementary yet long and elaborate; some are advanced and too compact and hence become difficult to grasp for our undergraduate students. Some contain some additional topics not included in the syllabus. Therefore the author felt the need for writing a book that was lucid and simple, neither too compact nor too elementary, that would contain the basics of a first course exactly as per syllabus with a number of examples and applications to illustrate the concepts.

Chapter 1–9 are the topics as per the UGC model syllabus and the syllabi of the Universities of Odisha including Biju Patnaik University of Technology. Chapter 10 deals with the rudiments of fuzzy sets which is meant for advanced and interested students.

The prerequisite of this book is only a course on higher secondary mathematics. However, efforts have been made to make the book self-contained with preliminary chapters on mathematical logic and set theory which are also included in the syllabus. This course is a prerequisite for advanced courses in various areas of computer science. Therefore this course is taught as a compulsory (core) course to BE/B Tech students all over the world and MCA students.

Besides mathematical logic and set theory, this book contains subjects like Boolean algebra, graph theory, counting, combinatorics and algebraic structures such as groups and rings-all these subjects are required for the study of computer science and hence are included in the syllabus for B.Tech Computer Science and Engineering. The contents of this book have been taught by the author several times to undergraduate students of computer science at the Indian Institute of Technology, Kharagpur and Kalinga Institute of Industrial Technology (KIIT) Deemed to be University, Bhubaneswar.

During the preparation of the manuscript, the author greatly benefitted by the works of several authors, a list appears in the references. The author takes this opportunity to acknowledge his indebtedness to them. This manuscript (which is at par the UGC model syllabus, and the syllabus of Universities of Odisha) is prepared when I am at Kalinga Institute of Industrial Technology Deemed to be University, Bhubaneswar, I thank Dr. Srikumar Acharya for editing, formatting and proof reading of this manuscript and I express profound thanks to Professor Achyuta Samanta, Founder of KIIT and KISS and the authorities of KIIT for cooperation and help. Finally I thank Allied Publishers Pvt. Ltd. for publishing the manuscript. I have taken enough care to make the book free from errors, but despite my efforts, some mistakes may have survived. I will be grateful to the readers to be kind enough to point them out. Finally, I must express that the purpose will be best served if it benefits students for whom it is meant.

Sudarsan Nanda

Dedicated

to

the memories of
my maternal grandparents
Kshetramani and Upendra Nath Mishra,
my parents Tilotama and Maheswar Nanda,
Kaviraj Lambodar Das,
my teachers Professors Ramanath Mohanty,
Brajabandhu Mishra and Sribatsa Nanda.

Contents

CHAPTER 1

Mathematical Logic

Logic deals with various kinds of reasoning based upon a set of hypotheses. It helps to increase the ability to communicate any idea in a precise way. So the rules (hypotheses), called the rule of inference, must be stated clearly in the most general form without ambiguity and must be independent of any specific discipline and any language used to describe them. Logic demands precision, but natural languages used by human beings are not precise enough. Therefore, in order to study logic, we first develop a formal language, called the object language, which consists of some well-defined terms and well-specified uses of these terms. Every branch of science uses its own object language, but in the study of logic, besides using it, we also analyse the object language. Further, in order to express the object language, we use another natural language and that is English for us. Logic is studied through the use of some symbols, which are explained by the object languages and hence, this branch of study is also called the symbolic logic.

1.1 CONNECTIVES

Statements

A statement is a declarative sentence which has one and only one of the two possible values, called the 'truth value'. These two truth values are 'true' and 'false', denoted by the symbols T and F, respectively; sometimes these are also denoted by the symbols 1 and 0. We assume that object language contains such declarative sentences. Since we allow only two possible truth values in our study, this logic is called two-valued logic.

Sentences

May God bless you (a wish),
Wish you a happy journey (a wish),
Please wait (a request),
What are you doing? (an enquiry)

are not statements, since they do not declare a definite truth value T and F.

But, the sentences like

India is a country,

Bhubaneswar is the capital of Odisha,

Kolkata is an old city, 41 is a prime number, are statements.

'This statement is false' is not a statement because we cannot assign a definite truth value to it. If we assign the value *T*, then the sentence says that the 'statement is false', if we assign *F*, then the statement is true.' This example is what is called a semantic paradox.

The sentence like,

$101 + 10 = 110$

is a statement. It has a truth value, but the truth value depends on the context; if we take decimal system, the statement is false, but in binary system it is true. Even then we consider such sentences as statements.

In order to build up the symbolic logic we use some symbols. We use single letters such as *A, B, C, ...* to denote statements. Another important step is that we use some 'logical connectives' to connect several statements into a single statement and frame rules for examining its truth value. We shall discuss these connectives in the following section.

Connective (Negation, Conjunction, Disjunction)

We shall now discuss three most basic and fundamental connectives. Other connectives will be discussed subsequently. These are negation, conjunction and disjunction and related to English word *not, and, or* respectively. Before discussing them we shall first list them along with their symbols:

Connective English Word	*Name of Connective*	*Symbol*
Not	Negation	$\neg$
And	Conjunction	$\wedge$
Or	Disjunction	$\vee$

Negation

Consider the statement

P: Kolkata is a city

The negation of *P*, $\neg P$ can be represented by

$\neg P$: Kolkata is not a city.

Since, in a natural language we write one statement in several ways, $\neg P$ can also be written as,

'It is not true that Kolkata is a city' or

'It is not the case that Kolkata is a city'.

Similarly, if

$\neg P$: I went to my office yesterday,

Then

P : I did not go to my office yesterday.

The following is the truth table for negation.

Truth Table for Negation

P	$\neg P$
T	F
F	T

Conjunction

The conjunction of two statements P and Q is read as 'P and Q' and is denoted by $P \wedge Q$. For example, if:

P : It is raining now

Q : There are 30 chairs in my room, then

$P \wedge Q$: It is raining now and there are 30 chairs in my room.

As another example, if:

P : Ram went to the school

Q : Gopal went to the school, then

$P \wedge Q$: Ram and Gopal went to the school.

The truth table is given as follows:

Truth Table for Conjunction

P	Q	$P \wedge Q$
T	T	T
T	F	F
F	T	F
F	F	F

Disjunction

The disjunction of two statements P and Q is the statement which is read as P or Q and is denoted by $P \vee Q$.

For example, if

P : There is something wrong with teacher

Q : There is something wrong with the student, then

$P \vee Q$: There is something wrong with the teacher or with the student.

This 'or' is inclusive or, that is, there may be something wrong either with the teacher or with the student or with both.

Consider another example

P : I shall watch the game on television

Q : I shall go to the game

$P \vee Q$: I shall watch the game on television or go to the game.

This is exclusive 'or', both P and Q cannot happen together. The truth table is as follows:

Truth Table for Disjunction

P	Q	$P \vee Q$
T	T	T
T	F	T
F	T	T
F	F	F

The truth table for all the above three connectives can be put together in the following way:

P	Q	$\neg P$	$P \wedge Q$	$P \vee Q$
T	T	F	T	T
F	T	T	F	T
T	F	F	F	T
F	F	T	F	F

1.1.1 *Example*

Construct the truth table for:

(i) $P \wedge \neg P$, (ii) $P \vee \neg Q$, (iii) $(P \vee Q) \vee \neg P$

P	$\neg P$	$P \wedge \neg P$
T	F	F
F	T	F

(ii)

P	Q	$\neg Q$	$P \vee \neg Q$
T	T	F	T
T	F	T	T
F	T	F	F
F	F	T	T

(iii)

P	Q	$P \vee Q$	$\neg P$	$(P \vee Q) \vee \neg P$
T	T	T	F	T
T	F	T	F	T
F	T	T	T	T
F	F	F	T	T

1.2 CONDITIONAL AND BICONDITIONAL

Let P and Q be any two statements. Then the statement '$P \Rightarrow Q$' which is read as 'if P then Q' or P implies Q is called a conditional statement. The truth value of '$P \Rightarrow Q$' is F when Q has truth value F and P the truth value T; in all other cases '$P \Rightarrow Q$' has truth value T. P is called the antecedent (hypothesis) and Q the consequent (conclusion) in '$P \Rightarrow Q$'. The truth table for '$P \Rightarrow Q$' is given as follows:

P	Q	'$P \Rightarrow Q$'
T	T	T
T	F	F
F	T	T
F	F	T

For any two statements P and Q, the statement '$P \Leftrightarrow Q$' is called a biconditional. This is read as 'P if and only if Q' and abbreviated as 'P iff Q'. This is also called 'P is necessary and sufficient for Q'. '$P \Leftrightarrow Q$' has the value T whenever both P and Q have identical truth values. The table is given as follows:

P	Q	$P \Leftrightarrow Q$
T	T	T
T	F	F
F	T	F
F	F	T

For any two statements P and Q, The statement '$Q \Rightarrow P$' is called the converse of '$P \Rightarrow Q$', $\neg P \Rightarrow \neg Q$' is called the inverse of '$P \Rightarrow Q$' '$\neg Q$' $\Rightarrow \neg P$' is called the contrapositive of '$P \Rightarrow Q$'. From the following truth table it will be clear that the original statement '$P \Rightarrow Q$' and the contrapositive '$\neg Q \Rightarrow \neg P$' have the same truth values.

1.2.1 *Example*

	Q	$P \Rightarrow Q$	$Q \Rightarrow P$	$\neg P$	$\neg Q$	$\neg P \Rightarrow \neg Q$	$\neg Q \Rightarrow \neg P$
T	T	T	F	F	F	T	T
T	F	F	T	F	T	T	F
F	T	T	F	T	F	F	T
F	F	T	T	T	T	T	T

1.2.2 *Example*

P	Q	$P \Rightarrow Q$	$Q \Rightarrow P$	$P \Leftrightarrow Q$
T	T	T	T	T
T	F	F	T	F
F	T	T	F	F
F	F	T	T	T

1.2.3 *Example*

There are two restaurants next to each other. One has a sign that says 'Good food is not cheap' and the other has a sign that says 'Cheap food is not good.' Are both the signs saying the same things?

Ans: Let G denote the statement that the food is good and C, the statement that the food is cheap. Hence the first sign says '$G \Rightarrow \neg C$' and the second says '$C \Rightarrow \neg G$'. It will now be clear from the following truth table that one is the contrapositive of the other and hence, they mean the same thing.

G	C	$\neg G$	$\neg C$	$G \Rightarrow \neg C$	$C \Rightarrow \neg G$
F	F	T	T	T	T
F	T	T	F	T	T
T	F	F	T	T	T
T	T	F	F	F	F

1.2.4 *Example*

Construct the truth table for $\neg (P \wedge Q) \Leftrightarrow (\neg P \vee \neg Q)$.

P	Q	$P \wedge Q$	$\neg (P \wedge Q)$	$\neg P$	$\neg Q$	$\neg P \vee \neg Q$	$\neg (P \wedge Q) \Leftrightarrow (P \vee \neg Q)$
T	T	T	F	F	F	F	F
T	F	F	T	F	T	T	T
F	T	F	T	T	F	T	T
F	F	F	T	T	T	T	T

Propositional Laws

We now mention the following propositional laws which are true for the logical connectives:

Commutative Laws : $(A \vee B) \Leftrightarrow (B \vee A)$, $(A \wedge B) \Leftrightarrow (B \wedge A)$

Associative Laws : $A \vee (B \vee C) \Leftrightarrow (A \vee B) \vee C$

$A \wedge (B \wedge C) \Leftrightarrow (A \wedge B) \wedge C$

Distributive Laws : $A \wedge (B \wedge C) \Leftrightarrow (A \wedge B) \vee (A \wedge C)$

: $A \vee (B \wedge C) \Leftrightarrow (A \vee B) \wedge (A \vee C)$

Idempotent Laws : $(A \wedge A) \Leftrightarrow \mathrm{A}$, $(A \vee A) \Leftrightarrow A$

Absorption Laws : $A \vee (A \wedge B) \Leftrightarrow A$, $A \wedge (A \vee B) \Leftrightarrow A$

De Morgan's Laws : $\neg (A \vee B) \Leftrightarrow (\neg A \wedge \neg B)$

$\neg (A \wedge B) \Leftrightarrow (\neg A \vee \neg B)$.

Some more important implications and equivalences are mentioned below:

Detachment : $(A \Rightarrow B) \wedge A) \Rightarrow B$

Disjunctive addition : $A \Rightarrow (A \vee B)$

Conjunctive simplification : $(A \wedge B) \Rightarrow A$

Chain rule : $((A \Rightarrow B) \wedge (B \Rightarrow C)) \Rightarrow (A \Rightarrow C)$

Disjunctive simplification : $(A \vee B) \wedge (\neg A) \Rightarrow B$

Conditional rule : $(A \Rightarrow B) \Leftrightarrow (\neg A) \vee B)$

$(A \Rightarrow B) \Leftrightarrow \neg B \Rightarrow \neg A$

Equivalence rule : $(A \Leftrightarrow B) \Leftrightarrow ((A \Rightarrow B) \wedge (B \Rightarrow A))$

$(A \Leftrightarrow B) \Leftrightarrow (A \wedge B) \vee ((\neg A) \wedge (\neg B))$

$(\neg (A \Leftrightarrow B)) \Leftrightarrow (A \Leftrightarrow (\neg B))$

Implication rules : $(A \Rightarrow B) \Rightarrow ((A \wedge C) \Rightarrow (B \wedge C))$

$(A \Rightarrow B) \Rightarrow (A \vee C) \Rightarrow (B \wedge C)$

Contra positive inference : $((A \Rightarrow B) \wedge (\neg B)) \Rightarrow (\neg A)$

Well-Formed Formulas (WFF)

We shall first discuss about a statement formula. It should be noted that a statement formula is not a statement; however a statement can be obtained from it by replacing the variable by statements. In order to define a formula, we need certain concepts like language, string etc.

Language consists of few symbols given below:

1. Connectives (negation, disjunction, conjunction etc.)
2. Equality denoted by =,
3. Element-hood or belongingness denoted by '$\in$',
4. Parenthesis denoted by (),
5. Variables denoted by x, y, ...

A string is any succession of symbols of the language written one after the other.

We now quote the definition of a formula. A formula is also called well formed formula and is abbreviated as WFF.

A string is called a formula if it has the following properties:

1. each letter is a term,
2. if x and y are terms, then $x = y$ is a formula,
3. if x and y are terms, then $x \in y$ is a formula,
4. if P is a formula, then $\neg P$ is a formula,
5. if P and Q are formulae, then $P \vee Q$ is a formula,
6. the only formulae are those given by the rules (1) to (5).

Tautologies

We have already discussed the three logical connectives, $\neg$, $\wedge$ and $\vee$ and we shall discuss many other connectives in subsequent sections. The statements which do not contain any connectives are called atomic or simple statements; whereas the statements containing one or more simple statements and some connectives are called molecular or compound statements. We shall discuss about statement formula, well-formed formula (wff) in a precise manner in Section 4 while dealing with predicate calculus. For the moment we may say that statement formulas are compound statements derived from some statement variables. Statement formulas are not statements they become statements when the statement variables used to define them are replaced by definite statements. The truth values of the resulting statement (when definite statements are replaced for statement variables in a statement formula) depend on the truth values of the statements substitute for the variables. Thus, the truth table of a resulting statement is the summary of all its truth values of the statements of values to the variables appearing in the formula. Therefore, sometimes the truth values of the resulting statement may be T and sometimes F. There are some formulas whose truth values are always true or always false irrespective of the truth values assigned to the variable.

A statement whose truth value is always T is called a tautology and the one whose truth value is always false is called a contradiction. Note that the negation of tautology is a contradiction and that of a contradiction is a tautology.

As example, let P be any statement and consider the statement $P \vee \neg P$, and $P \wedge \neg P$. The truth tables are given below. Observe that $P \vee \neg P$ is a tautology and $P \wedge \neg P$ is a contradiction.

P	$\neg P$	$P \vee \neg P$	$P \wedge \neg P$
T	F	T	F
F	T	T	F

Duality Law

Two formulae A and A^* are said to be duals of each other if either of them can be obtained from the other by replacing $\wedge$ by $\vee$ and $\vee$ by $\wedge$. The two connectives are called dual to each other. If the formula A contains the special variables T or F, then the its dual A^* is obtained by replacing T by F and F by T in addition to the changes suggested above. In fact there is no loss of generality if we confine only to the symbols $\vee$, $\wedge$ and $\neg$, because we will see later that any formula containing any other connectives can be replaced by an equivalent formula containing only these three.

1.2.5 *Examples*

(a) The dual of $(P \vee Q) \wedge R$ is $(P \wedge Q) \vee R$

(b) The dual of $(P \wedge Q) \vee T$ is $(P \vee Q) \wedge F$

(c) The dual of $\neg\ (P \vee Q) \wedge (P \vee Q)$ $(\mathrm{P} \vee \neg\ (Q \wedge \neg S)$
Is $\neg\ (P \wedge Q) \vee (\mathrm{P} \vee \neg\ (Q \wedge \neg S)$.

Functionally Complete Sets of Connectives

We have so far discussed the connectives $\vee$, $\wedge$, $\neg$, $\Rightarrow$ and $\Leftrightarrow$. We shall also discuss some other connectives in section 1.3. In fact we now show in this section that all the connectives are not essential. Given any formula containing all the above connectives, one can write an equivalent formula which contains proper subsets of these connectives.

A set of connectives is called functionally complete if every formula can be expressed in terms of an equivalent formula containing the connectives from this set.

Observe that $P \Leftrightarrow Q$ is equivalent to:

$$(P \Rightarrow Q) \wedge (Q \Rightarrow P).$$

1.2.6 *Examples*

(a) Write an equivalent formula for $P \wedge (Q \Leftrightarrow R) \vee (R \Leftrightarrow P)$ which does not involve biconditional.

Ans: $[P \wedge (Q \Leftrightarrow R) \vee (R \Leftrightarrow P)] \Leftrightarrow [P \wedge ((Q \Leftrightarrow R) \wedge (R \Rightarrow Q)) \vee (R \Leftrightarrow P) \wedge (P \Leftrightarrow R)]$

(b) Write an equivalent formula for $P \wedge (Q \Leftrightarrow R)$ which is free from biconditional as well as conditional.

Ans: $P \wedge (Q \Leftrightarrow R) \Leftrightarrow P \wedge ((Q \Leftrightarrow R) \wedge (R \Rightarrow Q)) \Leftrightarrow P \wedge ((\neg Q \vee R) \wedge (\neg R \vee Q))$.

Given a formula which involves all the five connectives $\vee$, $\wedge$, $\neg$, $\Rightarrow$ and $\Leftrightarrow$, we can have an equivalent formula by replacing biconditional and then by replacing the conditional. Finally, we can replace either $\vee$ or $\wedge$. This is equivalent to say that the set of connectives $\{\wedge, \neg\}$ and $\{\vee, \neg\}$ are functionally complete. Further, observe that $\{\neg\}$, $\{\wedge\}$, $\{\vee\}$ $\{\wedge, \vee\}$ are not functionally complete.

1.3 SOME MORE CONNECTIVES

Some connectives have already been discussed in section 1.1. In this section we will discuss some more connectives. We have seen that all the connectives are not essential for describing statements. For any formula, there exists an equivalent formula which can be expressed with the help of connectives belonging to one of the functionally complete sets. In spite of this fact, we define some more connectives, because some of the formulas can be expressed in simpler forms by using these connectives. We now discuss the connectives called exclusive OR, NAND, NOR.

Exclusive OR, NAND, NOR

Let P and Q be any two formulas. The formula $P \overline{\vee} Q$, where $\overline{\vee}$ is called an exclusive *OR*, is true whenever either P or Q is true, but not both. The exclusive *OR* is also called disjunction and is given by the following truth table.

P	Q	$P \overline{\vee} Q$
T	T	F
T	F	T
F	T	T
F	F	F

Observe that the following equivalences are true.

$$P \overline{\vee} Q \Leftrightarrow Q \overline{\vee} P \text{ (symmetric)}$$

$$(P \overline{\vee} Q) \overline{\vee} R \Leftrightarrow P \overline{\vee} (Q \overline{\vee} R) \text{ (associative)}$$

$$P \wedge (Q \overline{\vee} R) \Leftrightarrow (P \wedge Q) \overline{\vee} (P \wedge R) \text{ (distributive)}$$

$$(P \overline{\vee} Q) \Leftrightarrow (P \wedge \neg Q) \vee (\neg P \wedge Q)$$

$$(P \overline{\vee} Q) \Leftrightarrow (P \Leftrightarrow Q).$$

Also, the following are true (prove!)

If $(P \overline{\vee} Q) \Leftrightarrow R$, then $(P \overline{\vee} R) \Leftrightarrow Q$ and $(Q \overline{\vee} R) \Leftrightarrow P$, and $P \overline{\vee} Q \overline{\vee} R$ is a contradiction.

Before defining the other connectives, note that the word NAND stands for the combination of NOT and AND. The connective NAND is denoted by ↑. If P and Q are any two formulas, then we define NAND of P and Q by,

$$P \uparrow Q \Leftrightarrow \neg (P \wedge Q).$$

Note that in the above paragraph NOT stands for negation and AND for the conjunction.

The word NOR is a combination of NOT and OR where NOT and OR respectively stand for negation and disjunction. The connective NOR is denoted by ↓ and for any two formulas P and Q it is defined by,

$$P \downarrow Q \Leftrightarrow \neg (P \wedge Q).$$

The connectives ↑ and ↓ have been defined with the help of $\wedge$, $\vee$ and $\neg$. Therefore any formula involving ↑ and ↓, has an equivalent formula which can be expressed in terms of $\wedge$, $\vee$ and $\neg$. Further, Observe that ↑ and ↓, are dual to each other. Therefore, in order to obtain the dual of formula which involves ↑ or ↓, we have to interchange ↑ and ↓, in addition to other changes mentioned earlier.

We shall now show that each one of the connectives ↑ and ↓ is functionally complete. For this purpose it is enough to show that the sets of connectives $\{\wedge, \neg\}$ and $\{\vee, \neg\}$ can be expressed either in terms of ↑ alone or in terms of ↓ alone. The following expresses $\neg$, $\wedge$ and $\vee$ in terms of ↑ alone.

$$P \uparrow P \Leftrightarrow \neg (P \wedge P) \Leftrightarrow \neg P \wedge \neg P \Leftrightarrow \neg P$$

$$(P \uparrow Q) \uparrow (P \uparrow Q) \Leftrightarrow \neg (P \uparrow Q) \Leftrightarrow P \wedge Q$$

$$(P \uparrow Q) \uparrow (Q \uparrow Q) \Leftrightarrow \neg P \uparrow \neg Q) \Leftrightarrow \neg (\neg P \wedge \neg Q) \Leftrightarrow P \vee Q.$$

The following gives $\neg$, $\wedge$ and $\vee$ in terms of ↓ alone.

$$P \downarrow P \Leftrightarrow \neg (P \vee P) \Leftrightarrow \neg P \vee \neg P \Leftrightarrow \neg P$$

$$(P \downarrow Q) \downarrow (P \downarrow Q) \Leftrightarrow \neg (P \downarrow Q) \Leftrightarrow P \vee Q$$

$$(P \downarrow P) \downarrow (Q \downarrow Q) \Leftrightarrow \neg P \downarrow \neg Q) \Leftrightarrow \neg P \downarrow \neg Q \Leftrightarrow P \wedge Q.$$

The following are some basic properties of NAND and NOR

$$P \uparrow Q \Leftrightarrow Q \uparrow P, P \downarrow Q \Leftrightarrow Q \downarrow P \text{ (commutative)}$$

$$P \uparrow (Q \uparrow R) \Leftrightarrow P \uparrow \neg (Q \wedge R) \Leftrightarrow \neg (P \wedge \neg (Q \wedge R)) \Leftrightarrow \neg P \vee (Q \wedge R)$$

$$(P \uparrow Q) \uparrow R \Leftrightarrow (P \wedge Q) \vee \neg R.$$

Observe that the connective ↑ is not associative.

Two-State Devices

The logic that we have been discussing is called two-valued or bi-valued logic, because in this frame work any statement is allowed to have one and exactly one of the two possible truth values T and F. These two values T or F are also written as 1 or 0, respectively. There are some electrical and mechanical devices which have one of the two possible configurations like 'on' and 'off'. Here 'on' will represent for T or 1 and 'off' for F or 0. Therefore, such devices are called two-state devices. We now discuss some simple examples of two-state devices. An electric switch with 'on' and 'off' is a two state device. A vacuum tube or a transistor is another example of a two-state device in which the current is either passing (conducting) or not passing (non-conducting). We can have general discussion of such devices by replacing the word 'switch' by word 'gate' to mean a device which permits or stops the flow of not necessarily the electric current, but any quantity that can go through the device.

Consider an electric lamp controlled by a mechanical switch. Let p denote the (input) switch and s the (output) indicator lamp. When p is open, no current flows in the circuit and the lamp s is 'off'. When p is closed, s is 'on'. Let the statements be denoted by,

P : The switch p is closed.
S : The lamp s is on.

This can be expressed as a truth table or as a table for input-output values in the following manner:

State of p	*State of s*
Closed	On
Open	Off

Or

$p(P)$	$s(S)$
1	1
0	0

From now onwards throughout section 1.3 we shall write 1, 0 for T, F.

Consider a circuit in which there are two switches p and q in series. The lamp in this case is turned on whenever both the switches p and q are closed. Let,

P : The switch p is closed,
Q : The switch q is closed
S : The light s is on.

From the following it will be clear that $P \wedge Q \Leftrightarrow S$.

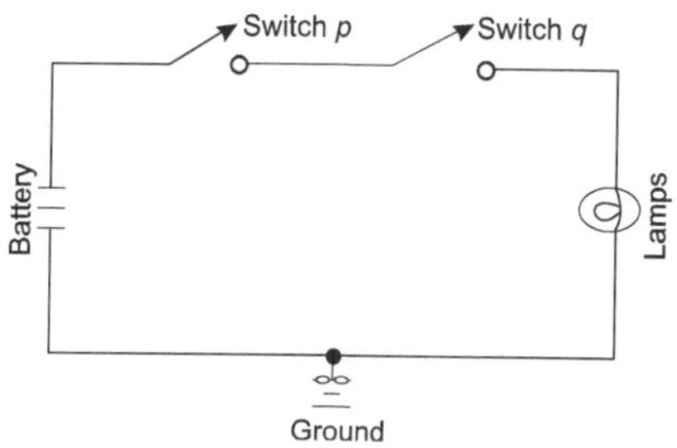

P	Q	S
1	1	1
1	0	0
0	1	0
0	0	0

The above figure shows a two-state device for AND logic.

The following figure and table show the case when two switches are connected in parallel. From the table it will follow that $P \vee Q \Leftrightarrow S$.

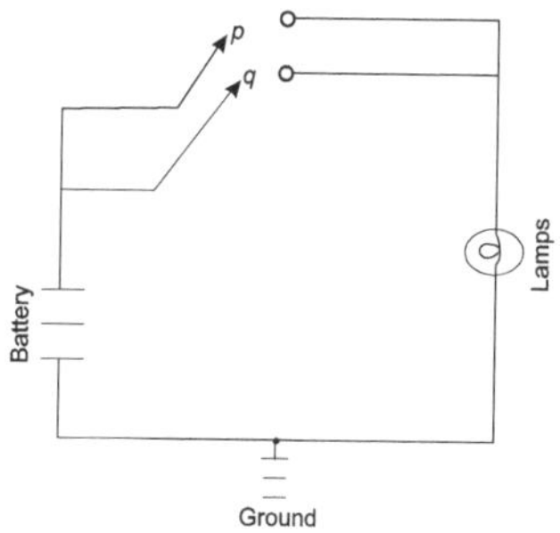

P	Q	S
1	1	1
1	0	1
0	1	1
0	0	0

The above figure shows a two-state device for OR logic.

Thus, Observe that the logical connections ∧ and ∨ correspond to switches connected in series and in parallel, respectively.

Gate and Module

We have just discussed that the logical connectives ∧ and ∨ can be considered as circuits (switches) connected in series and parallel respectively. Instead of representing ┐, ∨, ∧, by the circuits in the manner we have just discussed in the previous section or by some other equivalent circuit, they are usually represented by block diagrams, each such diagram having one or many input wires and one output wire. Such block diagrams are called gates.

In gates, the connectives ┐, v, and ∧ are usually denoted by the symbols –, + and – respectively. The block diagrams and tables for OR gate, AND gate and NOT gate are given respectively in the following diagrams.

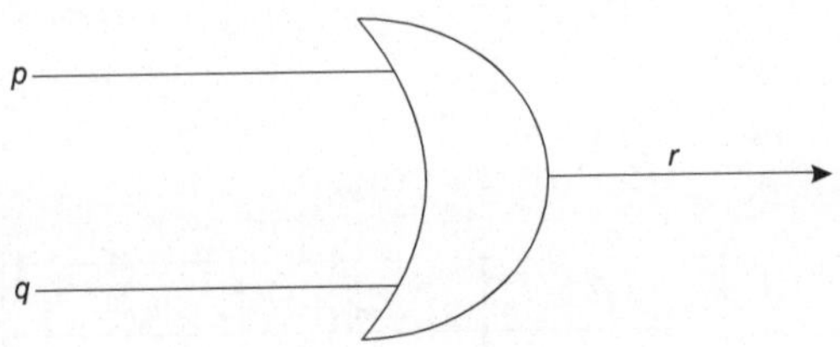

Input		*Output*
p	*q*	*r(p + q)*
1	1	1
1	0	1
0	1	1
0	0	0

An OR-gate

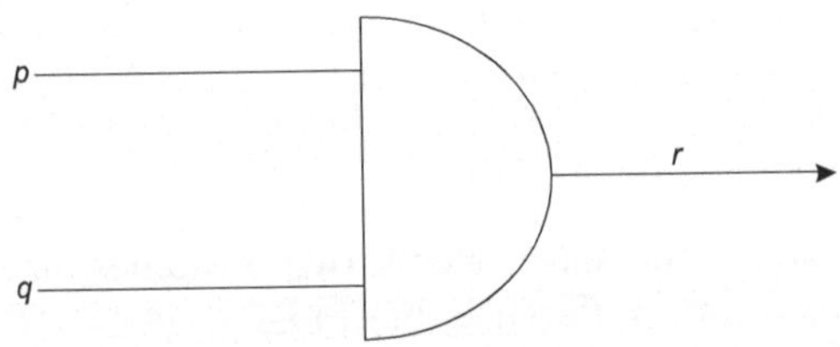

Input		*Output*
p	*q*	*r(p . q)*
1	1	1
1	0	0
0	1	0
0	0	0

An AND-gate

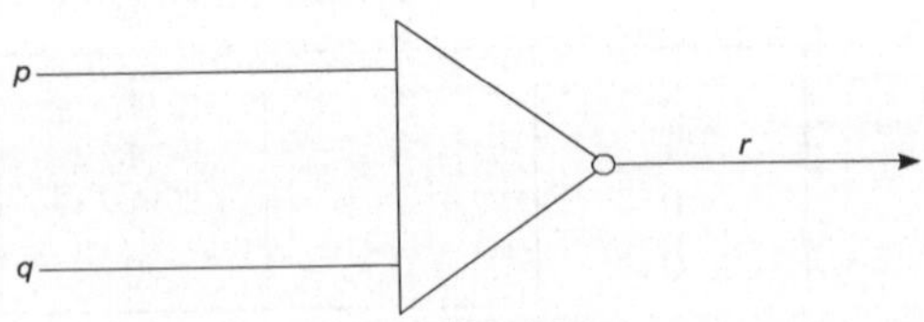

Input *q*	*Output* *r(p)*
1	0
0	1

A NOT-gate

Two-level Networks

A two-level network is defined as network in which the longest path through which information passes from input to output consists of two gates. A two-level network consisting of AND gates at the input stage followed by OR-gates at the output stage is called a sum-of product (AND-OR) network. An example of such a network is given by the expression $a \cdot b + c \cdot d$. This is shown by a table below.

Another type of two-level network is to have OR-gates at the input stage and then AND-gates at the output stages and such an expression is of the form $(a + b) \cdot (c + d)$. This is called a product-of-sum (OR-and) network and shown below.

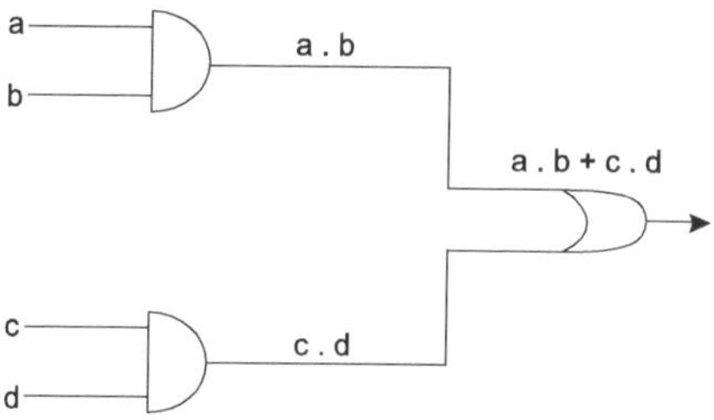

AND-OR

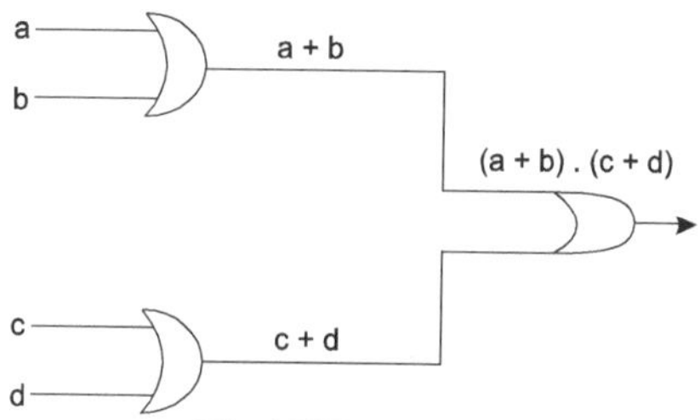

OR-AND

NOR and NAND Gates

There are two other types of gates which are useful and frequently used in computers. These are called NOR and NAND gates. The following tables show them, their tables and the equivalent representations:

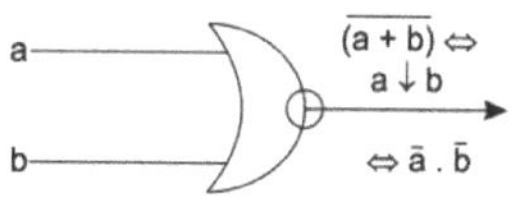

A NOR-gate

Input		*Output*
a	*b*	
1	1	0
1	0	0
0	1	0
0	0	1

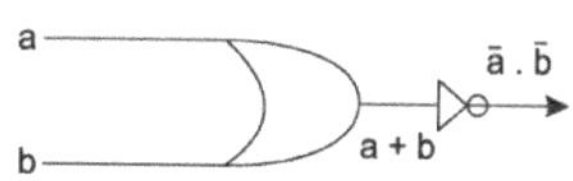

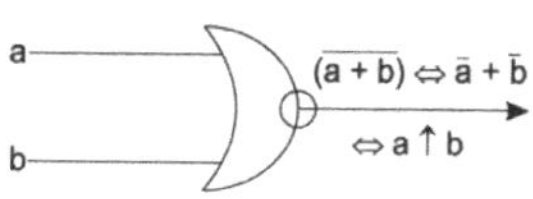

A NAND-gate

Input		*Output*
a	*b*	
1	1	0
1	0	1
0	1	1
0	0	1

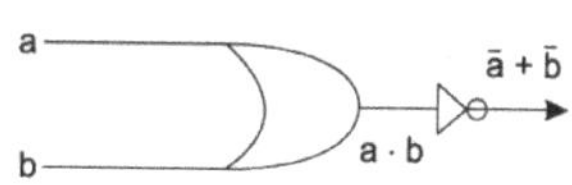

1.4 NORMAL FORMS AND THE THEORY OF INFERENCES

Disjunctive Normal Forms

In this section we shall use the words 'product' and 'sum' in place of the logical connectives conjunction and disjunction respectively. Before defining what is called 'disjunctive normal forms' we first introduce some other concepts needed in the sequel.

In a formula, product of the variables and their negations is called an elementary product. Similarly a sum of the variables and their negations is called an elementary sum.

Let P and Q be any two variables. Then $P, \neg P \wedge Q, \neg Q \wedge P \wedge \neg P, P \wedge \neg P, Q \wedge \neg P$ are some examples of elementary products; and $P, \neg P \vee Q, \neg Q \vee P \vee \neg P, P \vee \neg P, Q \vee \neg P$ are examples of elementary sum of two variables. A part of the elementary sum or product which is itself an elementary sum or product is called a factor of the original sum or product. The elementary sum or products satisfy the following properties. We shall only state them without proof.

1. An elementary product is identically false if and only if it contains at least one pair of factors in which one is the negation of the other.
2. An elementary sum is identically true if and only if it contains at least one pair of factors in which one is the negation of the other.

We now discuss some examples.

1.4.1 *Example*

Obtain the disjunctive normal forms of:

(i) $P \wedge (P \Rightarrow Q)$, (ii) $\neg P \vee Q \Leftrightarrow P \wedge Q$.

Proof

(i) $P \wedge (P \Rightarrow Q) \Leftrightarrow P \wedge (\neg P \vee Q) \Leftrightarrow (P \wedge \neg P) \vee (P \wedge Q)$

(ii) $(P \vee Q) \Leftrightarrow (P \wedge Q) \Leftrightarrow (\neg P (P \vee Q)) \wedge (P \wedge Q) \vee ((P \vee Q) \wedge \neg (P \wedge Q))$.

[Using '$R \Leftrightarrow S \Leftrightarrow (R \wedge$ '$S) \vee (\neg R \wedge \neg S)$]

$\Leftrightarrow (\neg P \wedge \neg Q \wedge P \wedge Q) \vee ((P \vee Q) \wedge (\neg P \wedge \neg Q))$

$\Leftrightarrow (\neg P \wedge \neg Q) \wedge (P \wedge Q) \vee ((P \vee Q) \wedge \neg P) \vee \neg Q) \vee ((P \vee Q \wedge \neg Q))$

$\Leftrightarrow (\neg P \wedge \neg Q) \wedge (P \wedge Q) \vee ((P \vee \neg P) \vee (Q \vee \neg P) \vee ((P \wedge \neg Q) \vee (Q \wedge \neg Q)$

It should be noted that the disjunctive normal form of a given formula is not unique. For example, consider $P \vee (Q \wedge R)$. This is already in disjunctive normal form. We can write,

$P \vee (Q \wedge R) \Leftrightarrow (P \vee Q) \wedge (P \vee R)$

$\Leftrightarrow (P \wedge P) \vee (P \wedge Q) \vee (P \wedge R) \vee (Q \wedge R)$

And this is another equivalent normal form. In subsequent sections we shall introduce the concepts principal normal forms which give unique normal form of a given formula.

Conjunctive Normal Forms

A formula which consists of a product of elementary sums and is equivalent to a given formula is called conjunctive normal form of the given formula.

1.4.2 *Example*

Obtain a conjunctive normal form of:

(i) $P \wedge (P \Rightarrow Q)$, (ii) $\neg P \vee Q \Leftrightarrow P \wedge Q$

Proof

(i) $P \wedge (P \Rightarrow Q) \Leftrightarrow P \wedge (\neg P \vee Q)$

(ii) $(P \vee Q) \Leftrightarrow (P \wedge Q) \Leftrightarrow \neg\ (P \vee Q) \Rightarrow (P \wedge Q)) \wedge ((P \wedge Q) \Rightarrow \neg\ (P \wedge Q))$.

[Using '$R \Leftrightarrow S$' $\Leftrightarrow (R \Rightarrow S) \wedge (S \Rightarrow R)$]

$\Leftrightarrow ((P \vee Q) \vee (P \wedge Q) \wedge (\neg\ (P \wedge Q) \vee (\neg P \wedge \neg Q))$

$\Leftrightarrow (P \vee Q \vee P) \wedge (P \vee Q \vee Q) \wedge ((\neg P \vee \neg Q) \vee (\neg P \wedge \neg Q))$

$\Leftrightarrow (P \vee Q \vee P) \wedge (P \vee Q \vee Q) \wedge (\neg P \vee \neg Q \vee \neg P) \wedge (\neg P \wedge \neg Q \vee \neg Q)$.

Principal Disjunctive Normal Forms

For two statement variables P and Q, construct all possible formulae which consist of conjunction of P or its negation and conjunctions of Q or its negation such that none of the formulae contain both a variable and its negation. Note that any formula which is obtained by commuting the formulas in the conjunction should not be included in the list because any such formula will be equivalent to one included in the list. For example, any one of $P \wedge Q$ or $Q \wedge P$ is included in the list, but not the both. For two variables P and Q there are $2^2 = 4$ formula included in the list. These are given by $P \wedge Q$, $P \wedge \neg Q$, $\neg P \wedge Q$, $\neg P \wedge \neg Q$.

These formulae are called *minterms.* One can construct truth table and observe that no two minterms are equivalent. The truth table is given by,

P	Q	$P \wedge Q$	$P \wedge \neg Q$	$\neg P \wedge Q$	$\neg P \wedge \neg Q$
T	T	T	F	F	F
T	F	F	T	F	F
F	T	F	F	T	F
F	F	F	F	F	T

Given formula are equivalent formula, an equivalent formula, consisting of minterms only is known as its principal disjunctive normal form or sum-of-products canonical form.

Principal Conjunctive Normal Forms

Given a number variables, *maxterm* of these variables is a formula which consists of disjunctions in which each variable or its negation but not both appear only once. Observe that the maxterms are the dual of minterms. Note that each of the minterms has the truth T for exactly one combination of the truth values of the variables and this fact can be seen from truth table. Therefore, each of the maxterms has the truth value F for exactly one combination of the truth of the variables. This fact can be derived from the duality principle (using the properties of minterms) or can be directly obtained from truth table.

For a given formula, an equivalent formula consisting of conjunctions of the maxterms only is known as its *principle conjunctive normal form or the product-of-sums canonical form.*

1.4.3 *Example*

Obtain the principals disjunctive normal form of:

(i) $P \Rightarrow Q$, (ii) $P \vee Q$, (iii) $\neg (P \wedge Q)$, (iv) $\neg P \vee Q$

Solution

(i) $P \Rightarrow Q \Leftrightarrow (P \wedge Q) \vee (\neg P \wedge Q) \vee (\neg P \wedge \neg Q)$

(ii) $P \vee Q \Leftrightarrow (P \wedge Q) \vee (P \wedge \neg Q) \vee (\neg P \wedge Q)$

(iii) $\neg P \wedge Q \Leftrightarrow (P(Q \vee \neg Q)) \vee (\neg P \wedge Q) \vee (\neg P \wedge \neg Q)$

(iv) $\neg P \wedge Q \Leftrightarrow (\neg P \wedge (Q \vee \neg Q)) \vee (Q \wedge (\neg P \wedge \neg P))$

$\Leftrightarrow \neg (P \wedge Q) \vee (\neg P \wedge \neg Q) \vee (Q \wedge P) \vee (Q \wedge \neg P)$

$\Leftrightarrow (\neg P \wedge Q) \vee (\neg P \wedge \neg Q) \vee (P \wedge Q).$

1.4.4 *Example*

Show the followings are equivalent formulas.

(i) $P \vee (P \wedge Q) \Leftrightarrow P$ (ii) $P \vee (\neg P \wedge Q)) \Leftrightarrow P \vee Q.$

Solution

(i) $P \Leftrightarrow P \wedge (Q \vee \neg Q) \Leftrightarrow (P \wedge Q) \vee (P \wedge \neg Q) P \vee (P \wedge Q)$

(ii) $\Leftrightarrow (P \wedge (Q \vee \neg Q)) \vee (P \wedge Q) \Leftrightarrow (P \wedge Q) \vee (P \wedge \neg Q)$

Valid Inference using Truth Tables and Direct Method of Proof

Let A and B be two statement formulas. It is said that 'B' follows from A' or 'B is a conclusion of the premise A' or B is a consequence of the premise A' if $A \Rightarrow B$ is a tautology.

Let $\{A_1, A_2, \ldots, A_n\}$ be a set of formulas. It is said that conclusion B follows from the set of premises $\{A_1, A_2, \ldots, A_n\}$ if $A_1 \wedge A_2 \wedge \ldots, \wedge A_n \Rightarrow B$ is a tautology.

Given a set of premises and a conclusion, it is possible, by using truth table, to know whether the conclusion follows from the given premises. But this method becomes tedious and time consuming when the number of variables present in all the formulas representing premises and conclusion are large. Also, there are situations where truth table technique is no longer available. Because of these facts, other methods have been developed and we shall discuss them in the following section.

Rules of Inference (Rules P and T)

We shall describe the method by which one derives that a formula is a consequence of set of given premises (also called hypotheses). We first quote two rules of inference called the rules of inference, the rules P and T.

Rule P : A premise may be introduced at any point of the derivation.

Rule T : A formula S may be introduced in a derivation if S is tautologically implied by any one or more of the preceding formulas in the derivation.

Some important implications and equivalences which will be used in the derivation are listed. All the implications and equivalences which will be listed now are not independent of each other; some of them can be derived from some others. So, one could start with minimum number of them, but we will not follow such an axiomatic scheme in this introductory text. Some of these were also listed earlier while discussing implications.

Implication

Simplification : $P \wedge Q \Rightarrow P$

$P \wedge Q \Rightarrow Q$

Addition : $P \Rightarrow P \vee Q$

$Q \Rightarrow P \vee Q$

$\neg P \Rightarrow$ '$P \Rightarrow Q$'

$Q \Rightarrow `P \Rightarrow Q'$

$\neg (P \Rightarrow Q) P \Rightarrow P$

$\neg (P \Rightarrow Q) P \Rightarrow Q$

$P, Q \Rightarrow P \wedge Q$

Disjunctive syllogism : $\neg P, P \vee Q \Rightarrow Q$

Modus ponens : $P, P \Rightarrow Q \Rightarrow Q$

Modus tollens : $\neg Q, P \Rightarrow Q \Rightarrow \neg P$

Hypothetical syllogism: $P \Rightarrow Q, Q \Rightarrow R \Rightarrow P \Rightarrow R$

Dilemma : $P \vee Q, P \Rightarrow R, Q \Rightarrow R \Rightarrow R$

Equivalences

Double negation : $\neg\neg P \Leftrightarrow P$

Commutative laws : $(P \wedge Q) \Leftrightarrow Q \wedge P, P \vee Q \Leftrightarrow Q \vee P$

Associative laws : $P \wedge Q \wedge R \Leftrightarrow P \wedge (Q \wedge R)$

$(P \vee Q) \vee R \Leftrightarrow P \vee (Q \vee R)$

Distributive laws : $P \wedge (P \vee R) \Leftrightarrow (P \wedge Q) \vee (P \wedge R)$

$P \vee (Q \wedge R) \Leftrightarrow (P \vee Q) \wedge (P \vee R)$

DeMorgan's laws : $\neg (P \wedge Q) \Leftrightarrow \neg P \wedge \neg Q$

$\neg (P \vee Q) \Leftrightarrow \neg P \wedge \neg Q$

$P \vee P \Leftrightarrow P, P \wedge P \Leftrightarrow P$

$R \vee (P \wedge \neg P) \Leftrightarrow R$

$R \wedge (P \vee \neg P) \Leftrightarrow R$

$R \vee (P \vee \neg P) \Leftrightarrow \mathrm{T}$

$R \wedge (P \wedge \neg P) \Leftrightarrow \mathrm{F}$

$P \Rightarrow Q \Leftrightarrow \neg P \vee Q$

$\neg (P \Rightarrow Q) \Leftrightarrow \neg P \wedge Q$

$P \Rightarrow Q \Leftrightarrow \neg Q \Rightarrow \neg P$

$P \Rightarrow (Q \Rightarrow R) \Leftrightarrow P \wedge Q \Rightarrow R$

$\neg (P \Rightarrow Q) \Leftrightarrow P \Leftrightarrow \neg Q$

$P \Leftrightarrow Q \Leftrightarrow (P \Rightarrow Q) \wedge (Q \Rightarrow P)$

$P \Leftrightarrow Q \Leftrightarrow (P \wedge Q) \vee (\neg P \wedge \neg Q)$.

Consistency of Premises and Indirect Method of Proof

If R is any formula, note that $R \wedge \neg R$ is a contradiction. A set of formulas $\{A_1, A_2, \ldots, A_m\}$ is said to be consistent if their conjunction has the truth value T for some assignment of the truth values to the atomic variables appearing in the formulas $\{A_1, A_2, \ldots, A_m\}$. Otherwise the formulas are called inconsistent. In other words, $\{A_1, A_2, \ldots, A_m\}$ are said to be inconsistent if every assignment of truth values to the atomic variables, at least one of the formulas $\{A_1, A_2, \ldots, A_m\}$ is false, i.e., their conjunction is identically false. To be precise, a set of formulas $\{A_1, A_2, \ldots, A_m\}$ is inconsistent if their conjunction is contradiction, i.e., $A_1 \wedge A_2 \wedge \ldots \wedge A_m \Rightarrow R \wedge \neg R$ where R is any formula.

The notion of inconsistency is used in a procedure called 'proof by contradiction or reduction absurdum or indirect method of proof'. In order to prove that a conclusion C follows from a set of premises $A_1, A_2, \ldots, A_m$, by using method 'proof by contradiction' the procedure is as follows:

Assume that C is false and consider $\neg C, A_1, A_2, \ldots, A_m$, is inconsistent they imply a contradiction, then the assumption that $\neg C$ is true doesn't hold simultaneously with $A_1 \wedge A_2 \wedge \ldots \wedge A_m$ being true. Therefore, C is true whenever $A_1 \wedge A_2 \wedge \ldots \wedge A_m$ is true. Thus the conclusion C follows from the premises $A_1 \wedge A_2 \wedge \ldots \wedge A_m$.

Proof by contradiction is sometimes very much convenient. But, if required, it can be eliminated and replaced by another method, called a conditional proof (CP). To demonstrate this observe that,

$$\text{`}P \Rightarrow (Q \wedge \neg Q) \Rightarrow \neg P\text{'} \quad \ldots (1)$$

In the proof by contradiction we show

$$A_1, A_2, \ldots, A_m \Rightarrow C$$

by showing

$$A_1, A_2, \ldots, A_m \neg C \Rightarrow (R \wedge \neg R).$$

Now this can be converted to the following by using rule *CP*

$$A_1, A_2, \ldots, A_m \Rightarrow \neg C \Rightarrow (R \wedge \neg R) \quad \ldots (2)$$

From (1) and (2) by using the rule of double negation one can prove that,

$$A_1, A_2, \ldots, A_m \Rightarrow C$$

And this is the required derivation.

Automatic Proving of Theorems

In this section we reformulate the rules of inference theory for statement calculus and describe a procedure of derivation which can be conducted mechanically.

Before describing this procedure, we first discuss some of the shortcomings of the procedure used in the earlier section. We shall then describe a set of 10 rules an axiom schema and rule for sequent's.

First of all observe that rule *P* permits the introduction of a premise at any point in the derivation, but it does not suggest either the premises or the step at which it should be introduced. Similarly, Rule *T* allows us to introduce any formula which follows from the previous step, but it does not suggest any definite choice of such formula nor there is any guidance for the use of any particular equivalence. Similarly, Rule CP does not tell anything about the stages at which an antecedent is to be introduced as an additional premise, nor does it indicate the stage at which it is again incorporated into the conditional. At every step, such decision is taken from a large number of alternatives, with the ultimate aim of reaching the conclusion.

We now describe our system which is given as follows:

1. *Variables* The capital letters of English alphabet *A, B, C, ... P, Q, R...* will be used to denote the statement variables. They are also used to represent the statement formulae, but these notations will be clear from the context.
2. *Connectives* The connectives $\neg$, $\wedge$, $\vee$, $\Rightarrow$, $\Leftrightarrow$ appear in the formula with order of precedence as given above.
3. *String of formulae A* string of formulae is defined as follows:
   (a) any formula is string of formulae,
   (b) if α and β are strings formula, then α, β and β, α are strings of formulae,
   (c) only those strings which are obtained from step (a) and (b) are strings of formulae; with the exception of the empty string which is also a string of formulae.

   The order in which the formulae appear in a string is not important, the strings *A, B, C*; *B, C, A*; *A, C, B*, etc. are the same.
4. *Sequents* if α and β are strings formulae then $\alpha \xrightarrow{S} \beta$ is called a sequent in which α is called the antecedent and β is called the consequent of the sequent.

   A sequent $\alpha \xrightarrow{S} \beta$ is true if either at least one of the formulae of the antecedent is false or at least one of the formulae of the consequent is true. The former case is called the vacuous proof and the latter is called the trivial proof.

Occasionally we have sequents which have empty strings of formulae as antecedent or as consequent. The empty antecedent, is interpreted as the logical constant 'true' or *T* and the empty consequent is interpreted as the logical constant 'false' or *F*.

5. *Axiom schema* if α and β are strings of formulae such that every formula in both α and β is a variable only, then the sequent $\alpha \overset{S}{\rightarrow} \beta$ is an axiom if α and β have at least one variable in common. For example, $A, B, C \overset{S}{\rightarrow} P, B, R$ are variable is an axiom.
6. *Theorem* the following sequents are theorems of our system.
   (a) every axiom is a theorem,
   (b) if a sequent α is a theorem and a sequent β results from α through the use of one of the 10 rules of the system which are given below, then β is a theorem.
   (c) Sequents obtained from (i) and (ii) are the only theorems.
7. *Rules:* The following rules are used to combine formulae within strings by introducing connectives. There are two rules corresponding to each of the connectives, one for the introduction of the connective in the antecedent and the other for its introduction in the consequent. While describing these rules, α, β, γ, … are used for strings of formulae while *X*, *Y* are used for formulae to which the connectives are applied.

Antecedent Rules

Rule ⇒ : If $\alpha, \beta \overset{S}{\Rightarrow} X, \gamma$, then $\alpha, X, \beta \overset{S}{\Rightarrow} \gamma$.

Rule ⇒ : If $X, Y, \alpha, \beta \overset{S}{\Rightarrow} \gamma$, then $\alpha, XY, \beta \overset{S}{\Rightarrow} X, \gamma$.

Rule ⇒ : If $X, \alpha, \beta \overset{S}{\Rightarrow} \gamma$; and $Y, \alpha, \beta \overset{S}{\Rightarrow} \gamma X, Y, \alpha, \beta \overset{S}{\Rightarrow} \gamma$, the $\alpha, XY, \beta \overset{S}{\Rightarrow} \gamma$.

Rule ⇒⇒ : If $Y, \alpha, \beta \overset{S}{\Rightarrow} \gamma\ \alpha, \beta \overset{S}{\Rightarrow} X, \gamma$, then $\alpha, X \Rightarrow Y, \beta \overset{S}{\Rightarrow} \gamma$.

Rule ⇒⇔ : If $X, Y, \alpha, \beta \overset{S}{\Rightarrow} \gamma, \alpha, \beta \overset{S}{\Rightarrow} X, Y, \gamma$, then $\alpha, X \Rightarrow Y, \beta \overset{S}{\Rightarrow} \gamma$.

Consequent Rules

Rule ⇒ : If $X, \alpha, \overset{S}{\Rightarrow} \beta, \gamma$, then $\alpha, \beta \overset{S}{\Rightarrow} \beta, X, \gamma$.

Rule ⇒ : If $\alpha, \overset{S}{\Rightarrow} X, \beta, \gamma, \alpha$ then $\alpha, \overset{S}{\Rightarrow} \beta, X, Y, \gamma$.

Rule $\Rightarrow$: If $\alpha, \overset{S}{\Rightarrow} X, Y, \beta, \gamma$ then $\alpha, \overset{S}{\Rightarrow} \beta, X, Y, \gamma$.

Rule $\Rightarrow\Rightarrow$: If $X, \alpha, \overset{S}{\Rightarrow} Y, \beta, \gamma$, then $\alpha, \overset{S}{\Rightarrow} \beta, X, \Rightarrow Y, \gamma$.

Rule $\Rightarrow\Leftrightarrow$: If $X, \alpha, \overset{S}{\Rightarrow} Y, \beta, \gamma$, and $Y, \alpha, \overset{S}{\Rightarrow} X, \beta, \gamma$, then $\alpha, \overset{S}{\Rightarrow} \beta, X, \Rightarrow Y, \gamma$.

1.5 THE PREDICATE CALCULUS

In our study of symbolic logic we have been discussing about 'statement calculus' so far. In statement calculus we consider statements as basic units, we do not analyse any atomic statement. Only analysis of compound formulas are done in statement calculus. Statements are also called propositions and 'statement calculus' is also called 'Propositional calculus'. Because we do not make any analysis of atomic statements in the study of statement calculus, it was not possible to compare any two atomic statements 'and also to know whether any two atomic statements have some features in common.

In order to study this aspect, we introduce the concept of a predicate in an atomic statement. The logic based upon the analysis of predicates in any statement is called predicate logic.

Predicates

We start with an example. Consider the two statements:

1. Ram is a bachelor.
2. Shyam is a bachelor.

In each statement 'is a bachelor' is a *Predicate.* These two statements are about two different persons, but both have some predicate 'is a bachelor'. In the study of statement calculus, we require two different symbols to express them and these symbols do not reveal the common property of these two statements. We now separate out the object (name of the individual) and the predicate. We introduce a symbol to denote the predicate 'is a bachelor' and a method to connect it with the names of persons. Then we will have symbols to denote the statements of persons being bachelor.

Consider another well known and commonly used example:

- All men are mortal.
- Ram is a man.
- Therefore, Ram is mortal.

Any human being will agree that the above conclusion is true so far as his intuition is concerned. But, it does not follow from the (inference theory of) statement calculus. In this case 'are mortal' or 'is mortal' is the predicate. When we separate out 'are mortal' from 'all men' and use different symbols for them, then it will be possible to draw conclusion about any particular man.

While constructing symbols, we shall use capital letters for the predicates and small letters for the objects. For example, let B denote the predicate 'is a bachelor', let r, s respectively stand for Ram and Shyam.

Then

$B(r)$: Ram is a bachelor.
$B(s)$: Shyam is a bachelor.

In general if a statement is given by 'P is Q' where Q is the predicate and P is the subject (object), then symbolically, it will be denoted by $Q(P)$.

Sometimes there is more than one subject associated with the same predicate. A predicate requiring m (m is a positive integer) objects is called an m-place predicate. If no names are associated with a predicate it is called a no-place predicate. Similarly 1-place predicate and 2-place predicate etc. are now self-explanatory.

Consider the following statements:

- Ram is younger than Shyam.
- Delhi is to the north of Hyderabad.

These are examples of 2-place predicates. In the first statement [is younger than' is the predicate, let us denote it by Y. As mentioned earlier, then r, s respectively denote Ram and Shyam. Then the first statement will be represented by $Y(r, s)$. Similarly in the second statement, let N denote the predicate 'to the north of', let d and h respectively denote Delhi and Hyderabad. Then the second statement will be denoted by $N(d, h)$.

Statement Functions, Variables and Quantifiers

Before giving the formal definition of a statement function, we first discuss an example. Let M denote the predicate 'is mortal'. Let r, d and t respectively denote Ram, Delhi and table. Observe that $M(r)$, $M(d)$ and $M(t)$ are Statements. Let $M(x)$ denote 'x is mortal'.

Note that $M(x)$ is not a statement. The statements $M(r)$, $M(d)$, $M(t)$ follow from $M(x)$ by replacing the variable x respectively by r, d and t. In general $M(x)$ will give us a statement when the variable x is replaced by an object. We now give a formal definition.

A simple statement function in one variable is an expression consisting of a predicate symbol and a single variable. Such a statement function becomes a statement when the variable is replaced by a definite object. The statement resulting from a replacement is called a substitution instance of the statement function and is a formula of statement calculus. A compound statement function is obtained from combining one or more simple statement functions and the logical connectives.

For example, if H(x) denotes 'x is a human being' and $M(x)$ denotes 'x is mortal'. Then the following are some examples of compound statement functions:

$H(x) \wedge M(x)$, $H(x) = M(x)$, $\rceil M(x)$,

$H(x) \vee \rceil M(x)$

In this section we shall discuss two other important concepts, namely the 'universal quantifier' and the 'existential quantifier'. We first discuss the universal quantifier. To this end we first start with some examples. The first set of examples will be from algebra. Consider

1. $x + 7 = 9$.
2. $x^2 - 2 = 0$.
3. $x^2 + 1 = 0$.
4. $(x - 2)\,(x - \frac{5}{6}) = 9$
5. $x^2 - 1 = (x - 1)\,(x + 1)$

In each of the above examples x is a variable. We get statements when x is replaced by definite objects. While studying elementary algebra, we do not like to replace x by any object like man or elephant or cabbage. We fix our universe from where the variable x should take values and the universe is the set N of natural numbers, Z of integers, Q of rational numbers, R of real numbers or C of complex numbers. Ex. 1 above is a true statement if x is replaced by 2. There is no rational number x for which Ex. 2 becomes a true statement. If we consider R as the universe, then there are two objects for which Ex. 2 is true. Ex. 3 has no truth value if x takes any real value; but in C, it becomes a true statement for two values of x. If we restrict to integers only Ex. 4 is a true statement for $x = 2$ only.

Example 5 is a true statement for each (every) value of x chosen from C.

Let us write Example 5 in the following way.

Example 6: For any number x, $x^2 - 1 = (x - 1)\,(x + 1)$.

Take another example.

Example 7: For any real number x, $x \geq 0$.

Note that Example 7 is a statement function and it becomes a (true) statement when the variable x is replaced by any number. But the situation in Ex. 6 is different, although it contains a variable x, it is not a statement function but is a statement. This happens because the phrase 'for any number x' appears there.

Actually if a statement like Ex. 6 is true for any variable x and involves an equality it is usually written as follows:

$x^2 - 1 \equiv (x - 1)(x + 1)$.

Consider another set of examples:

8. All men are mortal.
9. Every orange is yellow.
10. Any integer is either positive or negative.

We can write these examples as follows:

8'. For all x, if x is men, then x is mortal.
9'. For all x, if x is an orange, then x is yellow.
10'. For all x, if x is an integer, then x is either positive or negative.

We use a symbol to represent the phrase 'for all x', this is denoted by $(\forall x)$; some people also write it as (x). This is placed before the statement function to which it applies, thereby making it into a statement. This symbol is called the 'universl quantifier'. 'For all x' is also written as 'for every x' or 'for each x', etc.

Using this symbol we can rewrite the statements discussed above in the following manner. Let,

H(x) : x is man,
M(x) : x is mortal,
O(x) : x is an orange,
Y(x) : x is yellow,
N(x) : x is an integer,
P(x) : either, positive or negative.

Then Ex. 8', 9', 10' can be written as

8". $(\forall x)(H(x) \Rightarrow M(x)$
9". $(\forall x)(O(x) \Rightarrow Y(x))$
10". $(\forall x)(N(x) \Rightarrow P(x))$

Ex. 6, 7 can be written as,

6'. $(\forall x)\, x^2 - 1 = (x - 1)\,(x + 1)$.

7'. $(\forall x \in \mathrm{R}) x \geq 0$.

It is important to note that the particular variable appearing in the statements involving a universal quantifier is not important; the statements remain unchanged if the variable x is replaced by another variable y throughout. This agrees with our intuition. This is explained by an axiom of predicate calculus. As an example, we may mention here that 6' and 7' are respectively equivalent to 6" and 7" where,

6". $(\forall y) y^2 - 1 = y - 1)\,(y + 1)$.

7". $(\forall x \in R)\, y^2 \geq 0$.

Similarly 8", 9", 10" are respectively equivalent to 8"', 9"', 10"' where

8"'. $(\forall y)\,(\mathrm{H}(y) \Rightarrow \mathrm{M}(y)$

9"'. $(\forall y)\,(\mathrm{O}(y) \Rightarrow (\mathrm{Y}(y)$

10"'. $(\forall y)\,(\mathrm{N}(y) \Rightarrow \mathrm{P}(y))$

Another important point should be noted at this stage. Another axiom of predicate calculus expresses the idea that if a property P holds for every variable x belonging to a universal set and if C is a particular element of the same set, then P holds for C. By this axiom

All men are mortal,
Ram is a man,
gives the conclusion that
Ram is mortal.

We shall now discuss the other quantifier called the existential quantifier. For this purpose first consider the following examples.

11. There exists a man
12. Some men are clever.
13. Some real numbers are integers.

We can rewrite them as follows:

11'. There exists an x such that x is a man.

11". There exists at least one x such that x is a man.

12'. There exists an x such that x is a man and x is clever.

12". There exists at least one x such that x is a man and x is clever.

13'. There exists an x such that x is a real number and x is an integer.

13". There exists at least one x such that x is a real number and x is an integer.

We introduce a symbol denoted by $(\exists x)$, called the existential quantifier which symbolizes expressions such as 'there exists atleast one x such that or 'there exists an x such that' or 'for some x. Put

M(x) : x is man, C(x): x is clever,
R(x) : x is a real number,
I(x) : x is an integer.

Then the statements 11, 12, 13 can be expressed as follows:

11. $((\exists x)$ (M((x))
12. $(\exists x)$ (M(x) $\wedge$ C(x)
13. $((\exists x)$ (R(x) $\wedge$ $I(x)$).

Free and Bound Variables

We have discussed well-formed Formula (wff). We shall write wff by only a formula. We now discuss a formula containing a part of the form $\forall x$ P(x) or ($\exists x$ P(x). In this case languages will consist of the connectives $\forall$, $\exists$ besides the other connectives.

A letter x is said to be bound in a string S if $\forall x$ or $\exists x$ appears in S; otherwise x is said to be bound free. In other words, x is free in S if x does not appear in S.

A string containing a part of the form $\forall$, $\exists$ besides the other connectives is called a formula if it satisfies all the properties (i)–(v) described in 1.1 and the following properties: (v) if P is a formula and x is a letter free in $\exists x$ P, then $\forall x$ P is a formula, (vii) if P is a formula and x is a letter free in P, then $\exists x$ P is a formula, (viii) the only formulae are those given by the properties (i)–(vii).

In fact $\forall x$ $P(x) \wedge Q$(a) can be expressed in English as 'every x has the property P, and a has the property Q'. A change in variables in the bound occurrence is not a substitution instance. Sometimes it becomes necessary to change the variables to avoid confusion. It is better to write $\forall y$ $P(y) \wedge Q(x)$ instead of $\forall x$ $P(x) \wedge Q(x)$, so as to separate the free and bound variables. Occasionally, one may come across a formula of the type $\forall x$ $P(y)$ in which the occurrence of y is free and the scope of $\forall x$ does not contain x; in such a case, we have a vacuous use of $\forall x$.

It should be noted that in a statement every occurrence of a variable must be bound, and no variable should have a free occurrence. In the case where a free variable occurs in a formula, we have a statement function.

As examples, consider the following formulae:

$\forall x\, P(x, y)$ ... (1)

$\forall x\, (P(x) \Rightarrow Q(x))$ ... (2)

$\forall x\, (P(x)) \Rightarrow (\exists y)\, R(x, y)$ ... (3)

$\forall x\, (P(x) \Rightarrow R(x)) \vee \forall x\, (P(x) \Rightarrow Q(x))$ ... (4)

$(\exists x)\, (P(x) \wedge Q(x))$ ... (5)

$(\exists x)\, (P(x) \wedge Q(x)$ ... (6)

In (1) both occurrences of x are bound, while the variable y is free. In (2), all occurrences of x are bound. In (3) all occurrences of both x and y are bound, in (4) all occurrences of x are bound. In (6), the last occurrence of x in Q(x) is free.

Note that in the bound occurrence of a variable, the letter used to represent the variable is not important; in fact any other letter can be used as a variable without affecting the formula, provided that the new letter is not used elsewhere in the formula. Thus,

$\forall x\, P(x, y)$ and $\forall z\, P(z, y)$

are the same. It should also be noted that the bound occurrence of a variable cannot be substituted by a constant, only a free occurrence of a variable can be. For example,

$\forall x\, P(x) \wedge Q(a)$

is a substitution instance of:

$\forall x\, P(x) \wedge Q(y)$.

Special Valid Formulae Involving Quantifiers

Let A(x) be a predicate formula where x is an object variable of interest. Then

$\forall x\, P(x) \Rightarrow \mathrm{A}(y)$

where y is substituted for x in $A(x)$ to obtain A(y). We assume that $\forall x\, A(x)$ is true. Then $A(y)$ is true; If $\forall x\, A(x)$ is false, then nothing needs to be proved. This implication can be written in a more convenient form as,

$\forall x\, A(x) \Rightarrow A(x)$

This is called the rule of universal specification and is denoted by US. It states that from a universally valid premise $\forall x\, A(x)$, we can conclude a specific case $A(x)$, so that one can drop the quantifier in the derivation.

We shall now discuss another implication. For this, let x be any object variable, B be any formula not containing any free occurrence of x and $A(x)$ be any formula.

Then,

$$(B \Rightarrow A(x)) \Rightarrow (B \Rightarrow \forall x\, A(x) \quad \ldots (1)$$

This states that for an arbitrary object variable x if $A(x)$ follows from B, then we can conclude that $\forall x\, A(x)$ follows from B.

Note that if P, Q and R are Statement formulas, then

$$(P \vee \neg P) \Rightarrow Q) \Rightarrow (P \vee \neg P) \Rightarrow (Q \Rightarrow R) \quad \ldots (2)$$

As a special case of (1) let $B = P \vee \neg P$, where P is any statement variable. Since $P \vee \neg P$ is a tautology, from

$$(P \vee \neg P) \Rightarrow A(x) \Rightarrow (P \vee \neg P) \Rightarrow \forall x\, \mathrm{A}(x)$$

by using (2) we get,

$$Ax \Rightarrow \forall x\, \mathrm{A}(x).$$

This permits us to conclude $\forall x$ A(x) from A(x) so that we can add the universal quantifier in the conclusion during the course of derivation. This rule is called the rule of universal generalization and is denoted by UG.

Similar to US and UG, there are two more rules which permit us to remove or include the existential quantifier during the course of a derivation. They are

$$(\exists x)\, A(x) \Rightarrow A(y)$$

$$A(x) \Rightarrow (\exists x)\, A(x)$$

They are respectively called as existential specification (denoted by ES) and existential generalization (denoted by EG).

Theory of Inference for the Predicate Calculus

In the method of derivation in the predicate calculus, we use the rules of inference given for the statement calculus, the rules P and T. If the conclusion is given in the form of a conditional we also use the rule CP. Occasionally, we also use the indirect method of proof. Besides these, we also use the rules US, ES, UG and EG. These rules are described as follows. While describing the rules, Ax) will be used to denote a formula with a free occurrence of x. $A(y)$ will denote a formula obtained by the substitution of y for x in $A(x)$, for such a substitution $A(x)$ must be free from y.

- *Rule US:* From $\forall x\, A(x)$ one concludes $A(y)$.
- *Rule ES:* From $(\exists x)\, A(x)$ one can conclude $A(y)$ provided that y is not free in any given premise and also not free in any prior step of the derivation. These requirements can easily be met by choosing a new variable each time ES is used. (The conditions of ES are more restrictive than ordinarily required, but they do not affect the possibility of deriving any conclusion).
- *Rule EG:* From $A(x)$ one can conclude $((\exists y)\, x\, A(y)$.
- *Rule UG:* From $A(x)$ one can conclude $(\exists y)A(y)$ provided that x is not free in any of the given premises and provided that if x is free in a prior step which resulted from the use of ES, then no variables introduced by that use of ES appear free in $A(x)$.

EXERCISES

1. Construct the truth tables of the following propositions:
   (a) $(A \wedge B)\, v\, (\neg C)$
   (b) $(A \wedge B) \wedge ((\neg A) \vee (\neg C)$
   (c) $(\neg A) \vee (B)$
   (d) $A \wedge (\neg B)$
   (e) $\neg\, ((\neg A) \wedge (\neg B))$
   (f) $(\neg A) \vee (B \wedge (\neg C))$
   (g) $(A \wedge (\neg B)) \vee (C \wedge B)$
2. Construct the truth tables of the following propositions:
   (a) $A \wedge (\neg B) \Rightarrow ((\neg A)\, v\, (\neg C))$
   (b) $((A \wedge B) \Rightarrow (B \wedge C) \Rightarrow (B \wedge (\neg C)$
   (c) $(A \wedge B) \Rightarrow A,$
   (d) $(B \wedge (\neg A) \Rightarrow A$
   (e) $A \Rightarrow (A \vee B)$

CHAPTER 2

Set Theory and Integers

The axiomatic method of Mathematics which started with Euclid's Elements revived in the end of nineteenth century and made enormous progress since the beginning of twentieth century. Cantor (1845–1918) was the first to systematically study the theory of sets and hence, regarded as the creator or the father of set theory. He published his papers on Set Theory mostly during the period 1874–1897. Cantor tried to define a set. His definition was "A collection into a whole of definite, well distinguished objects of our intuition or thought". In his paper (3) Cantor believed.

"There exists a set *A* containing those and only those elements which satisfy any prepositional function ϕ".

The earlier attempt to axiomatize Cantor's Set Theory was made by *G.* Ferge. He included the so called axiom of abstraction which asserts the existence, for any given property *P*, of a set whose members are precisely those objects having the property.

Mathematicians did not work long with ideas of Cantor and Ferge before a crisis developed. One of the first and the simplest of such contradictions was made by Bertrand Russell in 1903 which is called Russell's Paradox.

This paradox led to various axiomatizations of Set Theory. The two types of collections: arbitrary collections in the sense of Cantor which we call *classes* and certain special type of collections that we call *sets* were distinguished. Russell and Whitehead [15] resolved this difficulty by discovering theory of types. They established a hierarchy of types of collections: *A* collection x can be a member of another collection y only if y is one level higher in the hierarchy than x. X is class if it is a collection of objects satisfying some prepositional function. Such a collection may not be a set.

Godel-Bernays system (see Bernays [1]) discussed the theory of classes which is assumed to have three primitives (undefined concept): sets, classes and membership. Zermelo-Frankel Theory (see Zermelo [20]) discussed set theory which is assumed to have two primitives: sets and membership. Thus, we do not define a set and an object or member of set.

2.1 SETS AND SET INCLUSION

We usually denote sets by capital letters such as *A*, *B*, *C*, ... and elements of a set by small letters such as *a*, *b*, *c*, ... To indicate that an element *a* is a member of a set *A*, we write,

$$a \in A$$

which is read '*a* is an element of *A*' or '*a* belongs to *A*'. To indicate that the element *a* is not a member of the set *A*, we write,

$$a \notin A$$

which is read '*a* is not an element of *A*' or '*a* does not belong to *A*'.

There are two standard notations available for designating a particular set. One way of describing a set is to list the elements of the set between braces. Thus, the set of all positive integers that are less than 5 can be written as,

$$\{1, 2, 3, 4\}$$

the set of all positive integers can be written as:

$$\{1, 2, 3, \ldots\}$$

the set of all odd integers can be written as:

$$\{\pm 1, \pm 3, \pm 5, \ldots.\}$$

and the set of all fourth roots of unity can be written as:

$$\{1, i, -1, -i\}$$

With this notation an arbitrary set with six members may be written as:

$$\{a, b, c, d, e, f\}$$

At times it is inconvenient or impossible to list all the elements of a set. In such cases, a set can often be described by specifying a property that the elements of the set have in common. For example, consider the set,

$$A = \{1, 2, 3, 4, 5, 6\}$$

which has been described here by listing the elements.

A is precisely the set of all positive integers that are less than 7 and this can be written as,

$$A = \{x \mid x \text{ is a positive integer less than } 7\}$$

which is read *A* is the set of all *x* such that *x* is a positive integer less than 7. As another example consider the following expression,

$$\{x \mid x \text{ is real and rational}\}$$

which is read 'the set of all x, such that x is real and rational, denotes the set of all real numbers which can be written as the quotient of two integers. The four sets of numbers described at the beginning of this paragraph can be written either way:

$$\{1,2,3,4\} = \{x|x \text{ is positive integer less than } 5\},$$
$$\{1,2,3,\ldots.\} = \{x\,|x \text{ is a positive integer}\},$$
$$\{1,i,-1,-\mathrm{i}\} = \{x\,|x \text{ is a complex number and } x^4 = 1\},$$
$$\{\pm 1, \pm 3,\ldots\} = \{x\,|x \text{ is an odd integer}\}.$$

Two sets A and B are equal if and only if they have the same elements. This is known as the *axiom of extension*. In this case we write,

$$A = B.$$

If A and B are not equal, then we write,

$$A \neq B.$$

For example, if

$$A = \{1,2,3\}$$

and,

$$B = \{x|x \text{ is a positive integer and } x^2 < 16\},$$

Then,

$$A = B.$$

Sometimes one sees sets that have no members. For example, the sets,

$$\{x\,|x \text{ is rational and } x^2 = 2\},$$
$$\{x\,|x \text{ is real and } x^2 = -1\}$$

have no members. A set with no members is called an *emptv* set or a null set and such a set is denoted by the symbol ϕ. A set is said to be *finite* if it is empty or consists of n elements for some positive integer n; otherwise it is said to be infinite.

More precisely the existence of the empty set follows from the *axiom of specification* which says that 'To every set A and to every condition $S(x)$ there corresponds a set B whose elements are exactly those elements x of A for which $S(x)$ holds'. By applying this axiom to any set A with the condition '$x \neq x$' we get the empty set,

$$\{x \in A : x \neq x).$$

If every element of a set A is also an element of set B, then we say that A is a subset of B or is contained in B and write,

$$A \subseteq B.$$

We sometimes express this by saying that B is a superset of A (or contains A). $A \subseteq B$ allows for the possibility that A and B might be equal. If A is a subset of B and A is not equal to B, then we say that A is a proper subset of B or is properly contained in B and this is written,

$$A \subset B.$$

We can also express this by saying that B is a proper superset of A or properly contains A. The existence of subsets follows from the axiom of specification. It is customary to agree that empty set ϕ is a subset of every set.

2.1.1 ***Example:*** If

$$A = \{a, b, c\}$$

then all possible subsets of A are,

$$\Phi, \{a\}, \{b\}, \{c\}, \{a, b\}, \{b, c\}, \{a, c\}, \{a, b, c\}.$$

2.1.2 ***Example:*** If,

$$A = (x \mid x \text{ is a positive integer}\},$$
$$B = \{x \mid x \text{ is an integer}\},$$

Then, $A \subset B$.

It is often convenient to have a symbol for logical implication. If p and q are statements, then $p \Rightarrow q$ means that p implies q, or that if p is true, then q is true. Similarly, $p \Leftrightarrow q$ means that p is equivalent to q, or that p is true if and only if q is true.

If A is any set, then,

$$A \subseteq A$$

Let A, B and C be three sets. Then,

$$A \subseteq B \text{ and } B \subseteq A \Rightarrow A = B,$$
$$A \subseteq B \text{ and } B \subseteq C \Rightarrow A \subseteq C.$$

Note that,

$$A = B \Leftrightarrow A \subseteq B \text{ and } B \subseteq A.$$

If a set A is not a subset of B we write,

$$A \not\subset B.$$

2.2 ALGEBRA OF SETS

The intersection of two sets A and B is the set of all elements that belong to both A and B, it is denoted by the symbol,

$$A \cap B.$$

If $A \cap B \neq \phi$ we say that A intersects B. If $A \cap B = \phi$, then we say that A does not intersect B or that A and B are disjoint. That $A \cap B$ is a set follows from the *axiom of specification* if we write,

$$A \cap B = [x \in A : x \in B\} \text{ or } A \cap B = \{x \in B : x \in A\}$$

The intersection of any collection of sets is the set of all elements that belong to every one of the sets in the collection.

2.2.1 ***Example:*** Let

$$A = \{1, 2, 4, 7, 11\}$$

$$B = \{2, 7, 9, 11, 13, 21\}$$

$$C = \{0, 2, 5, 7, 17, 23\}$$

$$D = \{2, 7, 13\}$$

Then

$$A \cap B \cap C \cap D = \{2, 7\}.$$

The operation of forming intersections is commutative and associative:

$$A \cap B = B \cap A$$

and

$$A \cap (B \cap C) = (A \cap B) \cap C.$$

It also has the property that,

$$A \cap A = A, A \cap \phi = \phi.$$

The union of two sets A and B is the set of all elements that belong to either A or B or both, it is denoted by the symbol,

$$A \cup B.$$

The above definition can also be expressed as,

$$A \cup B = \{x | x \in A \text{ or } x \in B\}.$$

The operation of forming unions is also commutative and associative,

$$A \cup B = B \cup A$$

and

$$A \cup (B \cup C) = (A \cup B) \cup C.$$

It also has the properties that,

$$A \cup A = A, A \cup \phi = A.$$

The union of any collection of sets is the set of all elements that belong to one or more of the sets in the collection. That it is a set follows from the

Axiom of union: For any collection of sets there is a set that contains all the elements that belong to at least one set of the given collection.

2.2.2 ***Example***

$$A = \{1, 2, 4, 7, 9, 17\}$$
$$B = \{2, 7, 9, 11\}$$
$$C = \{0, 17\}$$
$$D = \{1, 7, 13\}$$

Then,

$$A \cup B \cup C \cup D = \{0, 1, 2, 4, 7, 9, 11, 13, 17\}.$$

Note that the idea of using shading and circles to illustrate the relationship between sets is attributed to the British logician John Venn and in his honor, they are called Venn diagrams.

We have now defined two of the fundamental operations on sets and we shall now show that how they are related to each other. The facts here are given by the distributive laws:

$$A \cap (B \cup C) = (A \cap B) \cup (A \cap C)$$
$$A \cup (B \cap C) = (A \cup B) \cap (A \cup C)$$

We now give a logical proof of the distributive laws. We prove the first distributive law and leave the second as an exercise.

$$x \in A \cap (B \cup C)$$
$$\Leftrightarrow \quad x \in A \text{ and } x \in B \cup C$$
$$\Leftrightarrow \quad x \in A \text{ and } x \in B \text{ or } x \in A \text{ and } x \in C$$
$$\Leftrightarrow \quad x \in A \cap B \text{ or } x \in A \cap C$$
$$\Leftrightarrow \quad x \in (A \cap B) \cup (A \cap C)$$

Hence,

$$A \cap (B \cup C) = (A \cap B) \cup (A \cap C)$$

This proves the first distributive law.

The next of our major operations on sets is the formation of complements.

Given any set A, it may be necessary to consider those elements that are not in A. Usually, however, there are elements not in A that are extraneous to the problem under consideration. For example, let A be the set of all positive integers. Clearly, if x is a negative integer, then $x \notin A$. However, if x is a cabbage then x is also not in A since cabbages are not positive integers. If we are concerned with a problem about integers then 'an element x not in A' is really meant 'an integer x that is not in A'. Cabbages are extraneous to the problem.

One way of eliminating extraneous elements is to specify in advance that all elements in the problem under consideration come from some fixed set U, called the universal set. For example, in a problem concerned with real numbers, we might agree to take the universal set U to be the set of real numbers and by 'an element x that is not in A' we must mean 'a real number x that is not in A'. In many problems, the universal set is clear from the nature of the problem and is not explicitly stated.

If U is the universal set and A is a subset of U, then the set of all elements in U that are not members of A is called the complement of A; it is denoted by the symbol A'.

The operation of forming complements has the following obvious properties:

$$(A')' = A, U' = \phi, \phi' = U.$$

$$A \cup A' = U, A \cap A' = \phi.$$

2.2.3 ***Example:*** Let

$$U = \{0, 2, 9, 17, 23, 37\}$$

and

$$V = \{0, 17, 23\}$$

Then,

$$V' = \{2, 9, 37\}$$

The following two important results, called *De Morgan's laws,* give relationship of complementation with union and intersection.

If A and B are any two sets, then

$$(A \cup B)' = A' \cap B'$$

$$(A \cap B)' = A' \cup B'.$$

2.3.4 ***Example:*** To illustrate De Morgan's law,

Let,

$$U = \{a, b, c, d, e, f\}$$

$$A = \{a, b, d\}, B = \{b, d, f\}$$

Then,

$$A \cup B = \{a, b, d, f\}$$

$$(A \cup B)' = \{c, e\}$$

$$A' = \{c, e, f\}, B' = \{a, c, e\}$$

$$A' \cap B' = \{c, e\}.$$

We now prove the first De Morgan's law

$$x \in A' \cap B'$$

$\Leftrightarrow$ $x \in A'$ and $x \in B'$

$\Leftrightarrow$ $x \notin A$ and $x \notin B$

$\Leftrightarrow$ $x \notin A \cup B$

$\Leftrightarrow$ $x \in (A \cup B)'$

Hence,

$$A' \cap B' = (A \cup B)'$$

This completes the proof.

We now introduce the difference and symmetric difference of two sets. If A and B are two sets we define,

$$A - B = \{x \in A : x \notin B\}$$

That this is a set follows from the axiom of specification. It is easy to see that,

$$A - B = A \cap B'.$$

We define the symmetric difference of two sets A and B as,

$$A \Delta B = (A - B) \cup (B - A)$$

This operation is commutative and associative, i.e.

$$A \Delta B = B \Delta A$$

$$A \Delta (B \Delta C) = (A \Delta B) \Delta C.$$

Also,

$$A \Delta \phi = A, A \Delta A = \phi$$

It can be checked easily that,

$$A \Delta B = (A \cup B) - (A \cap B).$$

Given a set U and an element u if we use the digits 0 and 1 to denote $u \notin U$ and $u \in U$ respectively, then we can represent membership in tabular forms in the following manner,

Membership Table for **A'**	
A	A'
0	1

Membership Table for $A \cup B$		
A	B	$A \cup B$
0	0	0
0	1	1
1	0	1
1	1	1

Membership Table for $A \cap B$		
A	B	$A \cap B$
0	0	0
0	1	0
1	0	0
1	1	1

We have been considering the subsets of a given universal set U; do those subsets themselves constitute a set? The following axiom guarantees that the answer is yes.

***Axiom of Powers*:** For each set there exists a collection of sets that contains among its elements all the subsets of the given set.

In other words if U is a set, then there exists a set $P(U)$ such that if $X \subset U$, then $X \in P(U)$.

The set P described above may be larger than wanted; it may contain elements other than the subsets of U. This is easy to rectify; just apply the axiom of specification to form the set $\{X \in P(U) : X \subset U\}$. Since, for every X, a necessary and sufficient condition that X belongs to this set is that X be a subset of U, it follows that if we change notation and call this set $P(U)$ again, then,

$$P(U) = \{X: X \subseteq U\}.$$

This set $P(U)$ is called the power set of U; the axiom of extension guarantees its uniqueness.

Because the set $P(U)$ is very big in comparison with U, it is not easy to give examples. If $U = \phi$ then,

$$P(U) = \{\phi\}, P(\{a\}) = \{\phi, \{a\}\} \text{ and}$$

$$P(\{a, b\}) = [\phi, \{a\}, \{b\}, \{a, b\}].$$

The power set of triple has eight elements. The reader can probably guess (and is hereby challenged to prove) the generalization that includes all these statements: the power set of a finite set with n elements has 2^n elements.

2.3 CARTESIAN PRODUCT AND RELATIONS

In many situations we are interested in paired data. An ordered pair (a, b) is a listing of two objects a and b in definite order. The element a is called the first entry and the element b is called the second entry of the ordered pair. Two ordered pairs are said to be equal if they list the same objects in the same order. Thus, for (a, b) and (c, d) to be equal, we must have $a = c$ and $b = d$.

Thus two ordered pairs (a, b) and (b, a), though involve the same elements, convey different information because of the order unless $a = b$.

We are now in a position to define the Cartesian product of sets. If A and B are two sets, the Cartesian product of A and B is the set of all ordered pairs (a, b) where $a \in A$ and $b \in B$, it is denoted by $A \times B$.

Thus,

$$A \times B = \{(a, b) | a \in A, b \in B\}$$

2.3.1 *Example:* If $\mathrm{A} = \{\mathrm{a, b, c}\}$

and $B = \{1, 2\}$

then $A \times B = \{(a, 1), (a, 2), (b, 1), (b, 2), (c, 1), (c, 2)\}$.

It is possible to define the Cartesian product of more than two sets. For example, if $A_1, A_2 \dots A_n$ are n sets, where n is a positive integer, then the Cartesian product,

$$A_1 \times A_2 \times \dots \times A_n$$

is defined by

$$A_1 \times A_2 \times \dots \times A_n = \{(a_1, \dots, a_n) | a_1 \in A_1, \dots a_n \in A_n\}.$$

A relation is a set of ordered pairs. Thus, a relation is any set which is a subset of the Cartesian product of two sets. Let X and Y be any two sets (which may be equal, but need not be) and let R be a relation from X into Y. Then $R \subset X \times Y$. We define the domain of R to be the set dom $R = \{x \in X: (x, y) \in R \text{ for some } y \in Y\}$ and the range of R to be the set range $R = \{y \in Y: (x, y) \in R \text{ for some } x \in X\}$. The symbol R^{-1} denotes the inverse of R: $R^{-1} = \{(y, x): (x, y) \in R\}$.

There are two types of relations which often arise in mathematics: equivalence relations and order relations.

Before discussing them we will cite some examples.

2.3.2 *Example:* A marketing research firm classifies a person according to the following criteria.

$Sex: m =$ male; $f =$ female

Highest level of education completed:

e = elementary school, h = high school, c = college, g = graduate school.

Let $S = \{m, f\}, L = \{e, h, c, g\}$.

$S \times L$ contains all the categories into which the population is classified. The classification (f, g) represents a female who has completed graduate school.

2.3.3 *Example:* A software firm provides the following three characteristics for each program that it sells:

Language: f = FORTRAN, p = PASCAL, L = LISP

Memory: 48 = 48000, 64 = 64000, 128 = 128000

Operating system: u = UNIX, c = CP/M

$L = \{f, p, l\}, M = \{48, 64, 128\}, O = \{u, c\}$.

$L \times M \times O$ contains all the categories that describe a program.

2.3.4 *Example:* Let

A = set of all possible inputs to a given computer program,
B = the set of all possible outcomes.

Define a relation R from A to B:

aRb if b is the output produced by the program when the input a is used.

2.3.5 *Example:* An airline services the five cities C_1, C_2, C_3, C_4, C_5. The following table gives the cost (in rupees) of going from C_i to C_j.

To \ From	C_1	C_2	C_3	C_4	C_5
C_1		1400	1000	15000	2000
C_2	1900		2000	1600	2200
C_3	1100	1800		1900	2500
C_4	1900	2000	1200		1500
C_5	2000	1000	2000	1500	

Let $A = \{c_1, c_2, c_3, c_4, c_5\}$. Define a relation on A: $c_i R c_j$ if the cost of going from c_i to c_j is less than or equal to Rs. 1800/-. Find R.

$$R = \{(c_1, c_2), (c_1, c_3), (c_1, c_4), (c_2, c_4), (c_3, c_1), (c_3, c_2), (c_4, c_3), (c_4, c_5), (c_5, c_2), (c_5, c_4)\}$$

2.3.6 ***Example:*** Let $A = B = R$. Define xRy if $\frac{x^2}{4} + \frac{y^2}{9} = 1$.

R consists of all the points on the ellipse $\frac{x^2}{4} + \frac{y^2}{9} = 1$.

Let $A_1, A_2, A_3, \dots A_n$ be arbitrary sets. An n-ary relation R among the sets $A_1, A_2, A_3, \dots A_n$ is defined as a subset of $A_1 \times A_2 \times, A_3 \times \dots \times A_n$. In other words an n-ary relation among $A_1, \dots A_n$ is a set of ordered n-tuples as $(a_1, \dots, a_n)$ such that $a_i \in A_i$ for $i = 1, 2, \dots n$.

2.3.7 ***Example:*** Let supplier $= \{s_1, s_2, s_3, s_4\}$ be the set of all suppliers of parts, Part $= \{P_1, P_2, P_3, P_4, P_5, P_6, P_7\}$ be the set of parts, Project $= \{j_1, j_2, j_3, j_4, j_5\}$ be the set of projects and Quatity be the set N of positive integers. We can write a table called supply in the following manner:

Supply

Supplier	*Part*	*Project*	*Quantity*
s_1	P2	j_5	5
s_1	P3	j_5	17
s_2	P3	j_3	9
s_2	P1	j_5	5
s_4	P1	j_1	4

Observe that,

$$\text{Supply} = \{(s_1, P_2, j_5, 5), (s_1, P_3, j_5, 17), (s_2, P_3, j_3, 9), (s_2, P_1 j_5, 5), (s_4, P_1, j_1 4)\}$$

is a 4-ary relation, it is a subset of Supplier × Part × Project × Quantity.

2.3.8 ***Example:*** Consider a Table Assemble on Part, Part and Quantity. Let Part = $(p_1, p_2, .., p_7)$ be the set of parts and Quantity be the set N of positive integers. Let the Table be as follows:

Part	*Part*	*Quantity*
p_1	p_5	9
p_2	p_5	7
p_3	p_5	2
p_2	p_6	12
p_3	p_6	3
p_4	p_7	1
p_6	p_7	1

An ordered triplet (p_i, p_j, c) in the Table Assemble means that part p_i is a subcomponent of p_j, it takes c units of part p_i to assemble each unit of part p_j. Thus, Assemble is a subset of Part × Part × Quantity.

2.4 EQUIVALENCE RELATION

Definition: Let X be a nonempty set. An equivalence relation '$\sim$' on X is a relation such that for $x, y, z \in X$ we have, ·

1. $x \sim x$ (reflexive);
2. $x \sim y => y \sim x$ (symmetric);
3. $x \sim y$ and $y \sim z => x \sim z$ (transitive)

An equivalence relation on X has an intimate connection with decompositions of X into nonempty subsets which fill it out and have no elements in common with one another.

A partition of X is a disjoint class $\{X\}$ of nonempty subsets of X whose union is the full set X itself.

In other words, a partition of X is the result of splitting it, or subdividing it, into nonempty subsets in such a way that each element of X belongs to one and only one of the given subsets.

We now show that each partition of X has associated with it a natural equivalence relation in X and conversely, a given equivalence relation in X determines a natural partition of X.

Let X be a nonempty set. Assume that a partition of X is given. We now associate with this partition a relation '$\sim$' in X. We say that $x \sim y$ if x and y belong to the same partition set. It is easy to see that $\sim$ defines an equivalence relation in X.

Conversely let $\sim$ be an equivalence relation in X. Let x be an arbitrary element of X. The set defined by $[x] = \{y : y \sim x\}$, which is the set of all elements of X which are equivalent to x, is called the equivalence set of x. We show that the class of all distinct equivalence sets forms a partition of X. By reflexivity $x \in [x]$ for each $x \in X$, so each equivalence set is nonempty and their union is X. It remains to show that any two equivalence sets $[x_1]$ and $[x_2]$ are either disjoint or identical. Suppose that $[x_1]$ and $[x_2]$ are not disjoint. To be precise, let $z \in [x_1] \cap [x_2]$. Now $z \sim x_1$ and $z \sim x_2$ and by symmetry $x_1 \sim z$. Let $y \in [x_1]$, so that $y \sim x_1$ now since $y \sim x_1$ and $x_1 \sim z$ transitivity shows that $y \sim z$. Since, $y \sim z$ and $z \sim x_2$ again transitivity shows that $y \sim x_2$ so $y \in [x_2]$. Since y is arbitrary element of $[x_1]$, this shows that $[x_1] \subseteq [x_2]$.

The same reason shows that $[x_2] \subseteq [x_1]$.

Thus, $[x_1] = [x_2]$ and this completes the proof.

We now consider several important and simple examples of equivalence relations.

2.4.1 ***Example:*** Let I be the set of all integers. For $a, b \in I$ we define $a \sim b \Leftrightarrow a = b$. Then it is evident that ~ defines an equivalence relation in the set I and each equivalence class consists of precisely one integer.

2.4.2 ***Example:*** Let D denote the set of all fractions, i.e. let D be the set of all symbols of the form a/b, where $a, b \in I$ and b is not zero. If a/b and c/d are two fractions, then we define a relation ~in D as follows:

$$\frac{a}{b} \sim \frac{c}{d} \Leftrightarrow ad = bc.$$

Then for each $\frac{a}{b}, \frac{c}{d}, \frac{e}{f} \in D$.

We have

1. $\frac{a}{b} \sim \frac{a}{b}$;
2. $\frac{a}{b} \sim \frac{c}{d} \Rightarrow \frac{c}{d} \sim \frac{a}{b}$;
3. $\frac{a}{b} \sim \frac{c}{d}$ and $\frac{c}{d} \sim \frac{e}{f} \Rightarrow \frac{a}{b} \sim \frac{e}{f}$.

Thus ~ defines an equivalence relation in D and each equivalence class of fractions is what we call a rational number.

Function

Let X and Y be any two sets (which may be equal but need not be). A function from X into Y is a rule f which assigns to each element $x \in X$ unique element $y \in Y$. Then which y corresponds in this way to a given x is usually written f(x), and is called the image of x *under* the rule f, or the value of f at the element x. A function f can also be viewed as a relation, that is, a subset of $X \times Y$in which $(x, y) \in f$and $(x, z) \in f$implies that $y = z$.

A function f from X in to Y *is* often called a mapping, or a map, or a transformation from X into Y. The X is called the domain of f, written as dom f, and all the values of f for all x in X is called its range and is written as rng$f$. A function whose range consists of just one element is called a constant function.

If f is a function from X into Y it is denoted by the symbol $f: X \to Y$. If it is clear from the context what the sets X and Y are, or if there is no need of specifying them explicitly, it is a common practice to identify the function $f: X \to Y$ by the rule f without mentioning the sets X and Y. If two definite sets X and Y are under discussion, then a mapping of X into Y is sometimes designated by,

$$x \to y$$

where, y is the image of x. For example, the function $f: R \to R$ where R denotes the set of all real numbers defined by,

$$f(x) = x^2$$

can be designated by,

$$x \to x^2$$

A function f is called an extension of a function g (and g is called a restriction of f) if the domain of f contains the domain of g and $f(x) = g(x)$ for each x in the domain of g. If $f: X \to Y$,

let $G_f = \{(x, y) \in X \times Y: y = f(x)\}$. This set G_f is called the graph of f. The graph G_f is related to the function f in the same sense that either one of them completely determines the other.

Consider a mapping $f: X \to Y$. Then clearly range of f is a subset of Y, that is rng $f \subseteq Y$. If $f: X \to Y$ is such that range of f equals Y, then we call f surjective, or a mapping of X onto Y. If any two distinct elements in X have distinct images under f, then we call f injective, or a 1–1 mapping from X into Y. If f is both injective and surjective, then we call f bijective, or a one-one correspondence from X onto Y. If $f: X \to Y$ is bijective, then we can define its inverse mapping $f^{-1}: Y \to X$ as follows: for each $y \in Y$ there exists a unique element $x \in X$ such that $f(x) = y$ since f is 1–1 and on to; we then define x to be $f^{-1}(y)$.

Let X be a nonempty set. The identity mapping on X is the mapping of X onto itself which leaves each element of X fixed.

Suppose that $f: X \to Y$ and $g: Y \to Z$ be any two mappings. The product or the composition of these two mappings, denoted by $gf: X \to Z$, is defined by $(gf)(x) = g(f(x))$. We observe that the two-mappings involved here are not arbitrary, for the set Y which contains the range of the first, equals the domain of the second. More generally, the product of two mappings is meaningful whenever the range of the first is contained in the domain of the second.

2.4.3 *Example:* $f: R \to R$ be denoted by,

$$y = f(x) = x^2 + 1$$

and $g: R \to [-1, 1]$ be defined by,

$$g(x) = \sin x.$$

Then $gf: R \to [-1, 1]$ is defined by

$$(gf)(x) = g(f(x)) = g(x^2 + 1) = \sin(x^2 + 1).$$

Consider an arbitrary mapping $f: X \to Y$. If $A \subseteq X$, then its image $f(A)$ is the subset of. Y defined by

$$f(A) = \{f(x) | x \in A\}.$$

Similarly, if $B \subseteq Y$, then its inverse image $f^{-1}(B)$ is the subset of X defined by,

$$f^{-1}(B) = \{x | f(x) \in B\}.$$

The first mapping has the following properties:

$$f(\phi) = \phi; f(X) \subseteq Y$$

$$A_1 \subseteq A_2 \Rightarrow f(A_1) \subseteq f(A_2)$$

$$f\left(\bigcup_{i \in I} A_i\right) = \bigcup_{i \in I} f(A_i)$$

$$f\left(\bigcap_{i \in I} A_i\right) \subseteq \bigcap_{i \in I} f(A_i).$$

The second one is much better behaved:

$$f^{-1}(\phi) = \phi, f^{-1}(Y) \subseteq X,$$

$$B_1 \subseteq B_2 \Rightarrow f^{-1}(B_1) \subseteq f^{-1}(B_2)$$

$$f^{-1}\left(\bigcup_{i \in I} B_i\right) = \bigcup_{i \in I} f^{-1}(B_i)$$

$$f^{-1}\left(\bigcap_{i \in I} B_i\right) = \bigcap_{i \in I} f^{-1}(B_i).$$

Countable Sets and Counting Elements in Finite Sets

Let A and B any two sets. If there is a function f from A onto B which is also one-one, then f is called a one-one correspondence between A and B. If there is a one-one correspondence between two sets A and B, then we say that A and B are numerically equivalent or that A and B have the same cardinal number and we write it as $A \sim B$.

The relation '$\sim$' of numerical equivalence between two sets is in fact an equivalence relation. It can be easily verified that:

1. $A \sim A$ for any set A, i.e., '$\sim$' is reflexive,
2. If $A \sim B$, then $B \sim A$, i.e., '$\sim$' is symmetric,
3. If $A \sim B$ and $B \sim C$, then $A \sim C$, i.e., '$\sim$' is transitive.

A set is called finite if it is either empty or is numerically equivalent to the set $\{1, 2, 3, \cdots, n\}$ for some natural number n. A set A is called infinite if it is not finite. The simplest and most common example of an infinite set is the set N of all natural numbers:

$$N = \{1, 2, 3, \cdots\}.$$

A set A is said to be countable if it is finite or numerically equivalent to N. An example of a countable set is the set A of all even positive integers. In this case the function $f : A \to N$ defined by,

$$f(n) = \frac{n}{2}$$

is a 1–1 correspondence. Another example is the set A of all integers. In this case a 1–1 correspondence $f : N \to A$ is given by,

$$f(n) = \begin{cases} \dfrac{n}{2} & (n \quad \text{even}), \\ -\dfrac{n-1}{2} & (n \quad \text{odd}). \end{cases}$$

Counting Elements in Finite Sets

Let A and B two finite sets and suppose we know the number of elements in A and B. We investigate the number of elements in $A \cup B$ and $A \times B$.

If S is a finite set, then we denote the number of elements in S by the symbol $n(S)$.

2.4.4 Example: Let

$$A = \{3, 7, 19, 52, 61\}$$
$$B = \{z \mid z^4 = 1\}$$
$$C = \phi$$
$$D = \{x \mid x \text{ is an integer}\}.$$

Thus, we have

$$n(A) = 5, \ n(B) = 4, \ n(C) = 0 \text{ and } n(D) \text{ is undefined.}$$

If A and B are disjoint finite sets, then the number of elements in $A \cup B$ can be obtained by adding the number of elements in A and number of elements in B together; in other words,

$$n(A \cup B) = n(A) + n(B) \text{ if } A \cap B = \phi$$

2.4.5 Example: If $A = \{2, 3, 5, 7\}$ and $B = \{4, 9, 11\}$ then $A \cup B = \{2, 3, 4, 5, 7, 9, 11\}$.

Therefore,

$$n(A \cup B) = 7$$

On the other hand,

$$n(A) = 4 \text{ and } n(B) = 3$$

So that,

$$n(A \cup B) = n(A) + n(B).$$

If A and B are not disjoint then we cannot simply add $n(A)$ and $n(B)$ to obtain $n(A \cup B)$ because we would be counting the elements in $A \cap B$ twice. Since, we want to count these elements once, we can correct for this duplication by subtracting $n(A \cap B)$ once. Thus, we obtain

$$n(A \cup B) = n(A) + n(B) - n(A \cap B).$$

2.4.6 Example: Let $A = \{a, b, c, d, e\}$ and $B = \{c, d, e, f, g, h\}$

so that,

$$A \cup B = \{a, b, c, d, e, f, g, h\} \text{ and } A \cap B = \{c, d, e\}$$

Therefore

$$n(A) = 5, \ n(B) = 6, \ n(A \cup B) = 8, \ n(A \cap B) = 3$$

We have

$$8 = n(A \cup B) = n(A) + n(B) - n(A \cap B) = 6 + 5 - 3.$$

The above formula can be extended to union of three sets. More precisely, if A, B and C are three finite sets, then,

$$n(A \cup B \cup C) = n(A) + n(B) + n(C) - n(A \cap B) - n(B \cap C) - n(A \cap C) + n(A \cap B \cap C)$$

2.4.7 ***Example:*** Let $A = \{a, b, c\}$, $B = \{c, d, e, f\}$, $C = \{a, f, g\}$ then

$$A \cup B \cup C = \{a, b, c, d, e, f, g\}$$
$$A \cap B = \{c\},$$
$$B \cap C = \{f\},$$
$$A \cap C = \{a\}$$

and $A \cap B \cap C = \phi$

Thus,

$$7 = n(A \cup B \cup C) = n(A) + n(B) + n(C) - n(A \cap B) - n(B \cap C) - n(A \cap C) + n(A \cap B \cap C)$$
$$= 3 + 4 + 3 - 1 - 1 - 1 + 0.$$

2.4.8 ***Example:*** There were 100 students in a class. In an examination 80 students passed in English, 85 in Mathematics and 73 in both. How many failed in English and Mathematics both?

Solution: Let A denote the set of all students who passed in English and B denote the set of all students who passed in Mathematics.

Then

$$n(A) = 80, \quad n(B) = 85 \text{ and } n(A \cap B) = 73.$$

Now,

$$n(A \cup B) = n(A) + n(B) - n(A \cap B) = 80 + 85 - 73 = 92.$$

2.4.9 ***Example:*** Suppose we have six computers with the following specifications:

Computer	*Floating Point Arithmetic Unit*	*Magnetic Disc Memory*	*Graphic Display Terminal*
I	Yes	Yes	No
II	Yes	Yes	Yes
III	No	No	No
IV	No	Yes	Yes
V	No	Yes	No
VI	No	Yes	Yes

How many of the six computers have one or more of the three kinds of hardware considered?

Solution: Let A_1, A_2, A_3 be the sets of computers with a floating-point arithmetic unit, magnetic disc memory and graphic display terminal respectively. Then

$$n(A_1)=2,\ n(A_2)=5,\ n(A_3)=3,$$

$$n(A_1\cap A_2)=2,\ n(A_1\cap A_3)=1,\ n(A_2\cap A_3)=3$$

and

$$n(A_1\cap A_2\cap A_3)=1.$$

Hence,

$$\begin{aligned} n(A_1\cup A_2\cup A_3) &= n(A_1)+n(A_2)+n(A_3)-n(A_1\cap A_2) \\ &\quad -n(A_1\cap A_3)-n(A_2\cap A_3) \\ &\quad +n(A_1\cap A_2\cap A_3) \\ &= 2+5+3-2-1-3+1=5. \end{aligned}$$

Therefore, five of the six computers have one or more of the three kinds of hardware considered.

2.4.10 *Example:* Out of 200 students, 50 of them take the course Discrete Mathematics, 140 of them take the course Computer Language and 24 of them take both. Since both courses have scheduled examinations for the following day, only students who are not in either one of these courses will be able to go to the party the night before. How many students will be at the party?

Solution: Let A_1 and A_2 be the sets of students who take Discrete Mathematics and Computer Language respectively. Then

$$n(A_1)=50,\quad n(A_2)=140,\quad n(A_1\cap A_2)=24.$$

Hence,

$$\begin{aligned} n(A_1\cup A_2) &= n(A_1)+n(A_2)-n(A_1\cap A_2) \\ &= 50+140-24=166. \end{aligned}$$

Therefore, the number of students who will be at the party is,

$$200-166=34.$$

2.4.11 *Example:* A computer company must hire 25 programmers to handle systems programming tasks and 40 programmers for application programming. Of those hired, 10 will be expected to perform tasks of each type. How many programmers must be hired?

Solution: Let A_1 and A_2 respectively be the sets of system programmers and application programmers hired. Then,

$$n(A_1) = 25,$$
$$n(A_2) = 40,$$
$$n(A_1 \cap A_2) = 10.$$

Hence,

$$\begin{aligned} n(A_1 \cup A_2) &= n(A_1) + n(A_2) - n(A_1 \cap A_2) \\ &= 25 + 40 - 10 \\ &= 55. \end{aligned}$$

Therefore, 55 programmers must be hired.

2.4.12 ***Example:*** Thirty cars were assembled in a factory. The options available were a radio, an air-conditioner and white-wall tires. It is known that 15 of the cars have radios, 8 of them have air-conditioners and 6 others have white-wall tires. Moreover, 3 of them have all the three options. At least how many cars do not have any options at all.

Solution: Let A_1, A_2, A_3 be the sets of cars with a radio, an air-conditioner and white-wall tires respectively. Now,

$$n(A_1) = 15,$$
$$n(A_2) = 8,$$
$$n(A_3) = 6$$

and $n(A_1 \cap A_2 \cap A_3) = 3$.

Hence,

$$\begin{aligned} n(A_1 \cup A_2 \cup A_3) &= n(A_1) + n(A_2) + n(A_3) - n(A_1 \cap A_2) \\ &\quad - n(A_1 \cap A_3) - n(A_2 \cap A_3) + n(A_1 \cap A_2 \cap A_3) \\ &= 15 + 8 + 6 - n(A_1 \cap A_2) - n(A_1 \cap A_3) - n(A_2 \cap A_3) + 3 \\ &= 32 - n(A_1 \cap A_2) - n(A_1 \cap A_3) - n(A_2 \cap A_3) \end{aligned}$$

But

$$n(A_1 \cap A_2) \geq n(A_1 \cap A_2 \cap A_3)$$
$$n(A_1 \cap A_3) \geq n(A_1 \cap A_2 \cap A_3)$$
$$n(A_2 \cap A_3) \geq n(A_1 \cap A_2 \cap A_3)$$

Hence,

$$n(A_1 \cup A_2 \cup A_3) \leq 32 - 3 - 3 - 3 = 23$$

Thus, there are atmost 23 cars that have one or more options. Therefore, there are at least 7 cars that do not have any option.

Let A and B be two finite sets with $n(A) = m$ and $n(B) = n$. Then the members of $A \times B$ can be listed in a table with m rows and n columns. To be precise, let

$$A = \{a_1, a_2, \cdots, a_n\}$$

and

$$B = \{b_1, b_2, \cdots, b_n\}$$

Then $A \times B$ can be listed in a table as follows:

$A_1/B \rightarrow$	b_1	b_2	....	b_n
a_1	(a_1, b_1)	(a_1, b_2)	....	(a_1, b_n)
a_2	(a_2, b_1)	(a_2, b_2)	....	(a_2, b_n)
$\vdots$	$\vdots$	$\vdots$		$\vdots$
a_m	(a_m, b_1)	(a_m, b_2)		(a_m, b_n)

So that $A \times B$ has mn members.

This yields the formula,

$$n(A \times B) = n(A)\,.\,n(B)$$

The above formula can be extended to Cartesian product of three or more sets. We have

$$n(A_1 \times A_2 \times \cdots \times A_m) = n(A_1)\,.\; n(A_2) \cdots n(A_m)$$

2.4.13 ***Example:*** A coin is tossed four times and the resulting sequence of heads and tails is recorded. How many different sequences are possible?

Solution: The different sequences can be regarded as the members of the set,

$$A \times A \times A \times A$$

where $A = \{h, t\}$

Thus, the number of different sequences is,

$$n(A \times A \times A \times A) = n(A)\,.\,n(A)\,.\,n(A)\,.\,n(A)$$
$$= 2 \times 2 \times 2 \times 2 = 16$$

Counting Techniques, Multiplication Principle

There is a general counting principle that we shall use repeatedly; this is called the Multiplication principle and can be stated as follows.

If there are m ways to make a decision D_1 and there are n ways to make a decision D_2 then there are mn ways to make both decisions D_1 and D_2.

To see how multiplication principle holds, let

$$A = \{a_1, a_2, \cdots, a_m\}$$

be the set of possibilities for the decision D_1 and let,

$$B = \{b_1, b_2, \cdots, b_n\}$$

be the set of possibilities for the decision D_2.

To make decision D_1 and then the decision D_2 amounts to selecting an element from the set A and then selecting an element from the set B. But these are just ordered pairs, that is, the elements of the Cartesian product $A \times B$. But $A \times B$ has mn elements and thus, there are mn ways to make the two decisions.

The multiplication principle can be extended to more than two decisions.

If there are n_1 ways to make a decision D_1, n_2 ways to make a decision D_2, n_3 ways to make a decision D_3 and so on, then there are $n_1 . n_2 . n_3 ....$ ways to make all the decisions.

2.4.14 ***Example:*** A coin tossed six times, generates sequence of heads and tails. How many different sequences are possible?

Solution: There are two possibilities, head or tail, on each of the six tosses. Thus, by the multiplication principle, there are $2^6 = 64$ different sequences.

2.4.15 ***Example:*** A telephone number consists of a sequence of seven digits (with repeatation allowed) selected from the set {0, 1, 2, 3, 4, 5, 6, 7, 8, 9}. How many different telephone numbers are possible?

Solution: There are 10 possibilities for each of the digits. Thus, by the multiplication principle there are 10^7 different possible telephone numbers.

2.5 PEANO AXIOMS

Oldest and most basic mathematical system is the set N of all positive integers. Kronecker once said "God made the positive integers; all else is due to man". However, it was shown by the Italian Mathematician G. Peano (1858–1932), that the positive integers can even more simply be defined by a set of axioms, known as Peano axioms. To describe these axioms we first need to define the so-called Peano's successor function. This function, denoted by s, assigns to every positive integer the next larger one. Thus $s : N \to N$ is defined by,

$$s(n) = n + 1.$$

In terms of the function s, a number of basic properties can be formulated which characterize the set N. These properties can be summarized as follows:

(a) $1 \in N$;

(b) $n \in N \Rightarrow s(n) \in N$;

(c) $\exists$ no $n \in N$ such that $s(n) = 1$;

(d) $s(n) = s(m) \Rightarrow n = m$;

(e) if $A \subset N$ such that $1 \in A$ and $s(n) \in A$ whenever $n \in A$, then $A = N$.

Axiom (*e*) leads to one of the most important principles of Mathematics, known as the principle of mathematical induction.

Integers

We write N for the set of all positive integers and Z for the set of all integers.

Given $a, b \in Z$ $(b \neq 0)$ if there is $c \in Z$ such that $a = bc$, then we say that b divides a and we write $b|a$. It is clear that:

(a) If $c|b$ and $b|a$, then $c|a$

(b) If $c|b$ and $c|a$, then for any $m, n \in Z$ we have $c|ma + nb$.

Positive integers can be divided into two classes: Prime numbers and Composite numbers.

Definition

$p \in N$ is said to be a prime number if:

(a) $p > 1$

(b) p is not divisible by any positive integer except 1 and p

On the other hand, any positive integer greater than 1 which is not a prime is called a composite number.

2.5.1 *Theorem:* Every positive integer greater than 1 is the product of (certain) prime numbers.

Proof: Suppose that the given number is c. If it is a prime number then there is nothing to prove. If it is a composite number then it is divisible by some integers greater than 1 and less than c. Let p_1 be the smallest of such integers. Then p_1 is a prime number. For if p_1 is divisible by some positive integer $l > 1 (l \neq p_1)$ then l will also divide c. But l being less than p_1 this contradicts the choice of p_1. Thus c is divisible by a prime number p_1. Let $c = n_1 p_1$. Here $1 < n_1 < c$. Now n_1 is either a prime or is divisible by a prime number p_2. i.e. $n_1 = n_2 p_2$ where $1 < n_2 < n_1$. Similarly n_2 is either prime or is divisible by a prime number p_3. Thus $n_2 = n_3 p_3$ where $1 < n_3 < n_2$. Continuing in this way we get a decreasing sequence $\{n_1, n_2, n_3, \ldots\}$ of positive integers each greater than 1 and less than c. Let n_k be the smallest of these. Now n_k is a prime number ($= p_k$) for otherwise if n_{k+1} divides n_k then n_k will not be the smallest of $\{n_1, n_2, n_3, \dots\}$. Hence $n = p_1, p_2, p_3, \dots p_k$ and this completes the proof.

Note that the prime numbers $p_1, p_2, p_3, \dots$ need not be different from each other. If some of the p_m's are identical, then c will be represented as $c = p_1^{a1} p_2^{a2} \dots p_m^{am}$ ($a_1 > 0, a_2 > 0, \dots, p_1 < p_2 < \dots$)).

We now prove.

2.5.2 *Theorem* (*The fundamental theorem of Arithmetic*)
The above representation of c is unique. Before proving this we first prove

2.5.3 *Theorem* *(Euclid's First Theorem)*
If some prime number p divides ab, then p divides either a or b.

Proof: Suppose that p does not divide a. We now show that p divides b.

Firstly, consider the set ($pX + aY$: X, Y are integers). Clearly, $pX + aY$ is an integer whenever X and Y are integers. Let q be the least positive integer which can be expressed in the above manner. So $q = px_1 + ay_1$ for some $x_1, y_1 \in Z$. For some integers X and Y if $pX + aY \in N$ then this is either greater than q or equal to q.

Let $pX + aY = qs + r = (px_1 + ay_1)s + r$

$$p(X - x_1 s) + a(Y - y_1 s) = r$$

Now r is of the form $pX + aY$. If r would be positive, then it must be greater than or equal to q. But clearly $r < q$. Therefore, $r = 0$. Thus any integer which can be expressed in the form $pX + aY$ is divisible by q. Also $p = p.1 + a.0$. Thus p is divisible by q. But p is prime number and hence $q = 1$ or $q = p$. Since q divides $a = p.0 + a.1$, q cannot be equal to p. So $q = 1$. Thus

$$px_1 + ay_1 = 1$$

$$px_1 b + ay_1 b = b$$

Since p divides both $px_1 b$ and $ay_1 b$, we conclude that p divides b.

Proof of Theorem 2.5.2: Let $p_1^{a_1} p_2^{a_2} \dots p_k^{a_k} = c = q_1^{b_1} q_2^{b_2} \dots q_j^{b_j}$. Since q_1 divides $p_1^{a_1} p_2^{a_2} \dots p_k^{a_k}$, q_1 must divide some p_1. Again, since q_1 and p_1 are prime numbers, $p_1 = q_1$. Similarly each p is equal to some q and each q is equal to some p. So $k = j$ and $p_1 = q_1, p_2 = q_2, \dots p_k = q_k$. If $a_1 < b_1$ then since,

$$p_2^{a_2} p_3^{a_3} \dots p_k^{a_k} = q_1^{a_1-b_1} q_2^{b_2} \dots q_k^{b_k}$$

q_1 will divide at least one $p_2, p_3, p_4, \dots, p_k$. So $a_1 < b_1$ and similarly $b_1 < a_1$ are impossible. Hence $a_1 = b_1$ and similarly, $a_2 = b_2$, $a_3 = b_3, \dots a_k = b_k$ and this completes the proof. Let a and b be two given positive integers.

Consider the set $\{ax + by: x \text{ and } y \text{ are integers}\}$. Such a set is called a *Number Ring*. Let d be the smallest positive integer which belongs to this set. Thus there are integers x_1 and y_1 such that $d = ax_1 + by_1$ and if $ax + by$ is a positive integer then this is either greater than or equal to d.

We now prove that d divides each element of the number ring and if some integer n divides a and b both, then n divides d.

Proof: Let $a = dq + r$ where $0 \le r < d$.

$$r = a - dq = a - (ax_1 + by_1)q = a(1 - qx_1) + b(-qy_1) = ax_2 + by_2$$

If r would be positive then it would be greater than or equal to d. But this being impossible $r = 0$. Thus d divides a and similarly d divides b.

If there is some integer n which divides a and b both then $a = nk$ and $b = nl$ for some integers k and l. So $d = ax_1 + by_1 = (kx_1 + by_1)u$ and hence $n|d$.

Note: d defined as above is said to be the *Highest Common Factor* (H.C.F) of a and b and is written as (a, b).

It is clear that if $d = (a, b)$ in order that the equation $ax + by = n$ will have integer solutions d must divide n. On the other hand if d divides n then we can find atleast one pair of integers (x, y) for which $ax + by = n$.

If $(a, b) = 1$ then a and b are said to be *co-prime* and in this case a and b do not have any common factor other than 1.

Similarly for given m positive integers $a_1, a_2, \cdots, a_m$ we can find a positive integer d such that $d$divides each a_k and if any other positive integer n divides each a_k then n must divide d. This d is called the H.C.F of $a_1, a_2, \cdots, a_m$ and written as $d = (a_1, a_2, \cdots, a_m)$. As before we can find m integers $x_1, x_2, \cdots, x_m$ for which

$$d = a_1x_1 + a_2x_2 + \cdots + a_mx_m$$

We now prove this by induction on m. Firstly note that the result is true for $m = 2$. If $m > 2$, then observe that $(a_1, a_2, \cdots a_m) = ((a_1, a_2, \cdots, a_{m-1}), a_m)$. For if $d_{m-1} = (a_1, a_2, \cdots, a_{m-1})$ and $d_m = (a_1, a_2, \cdots, a_m)$ then $d_m | a_i$ $(i = 1, 2, \cdots, m)$. So $d_m | d_{m-1})$, but $d_m | a_m$. On the other hand if $c | d_{m-1})$ and $c | a_m$ then $c | d_m$. So $d_m = (d_{m-1}, a_m)$.

Now assume that if $d_r = (a_1, a_2, \cdots, a_r)$ then there are integers $y_1, y_2, \cdots, y_r$ for which $d_r = a_1y_1 + a_2y_2 + \cdots + a_ry_r$ holds. Then for two integers z_{r+1} and w_r we have

$$\begin{aligned} d_{r+1} &= (d_r, a_{r+1}) = d_rw_r + a_{r+1}z_{r+1} \\ &= a_1(y_1w_r) + a_2(y_2w_r) + \cdots + a_r(y_rw_r) + a_{r+1}z_{r+1} \\ &= a_1z_1 + a_2z_2 + \cdots + a_rz_r + a_{r+1}z_{r+1} \end{aligned}$$

This completes the proof.

The H.C.F. of two natural numbers a and b is obtained by *Euclid's algorithm* which is described below.

Firstly if $a = b$ then $(a, b) = (a, a) = a$. Assume that $a \neq b$ and to be more specific let $b < a$. Divide a by b and let r_1 be the remainder. Then divide b by r_1 and let r_2 be the remainder. Now divide r_1 by r_2 and let r_3 be the remainder. Continuing in this way we get $r_4, r_5 \cdots$ such that $b > r_1 > r_2 > r_3 > \cdots r_k > r_{k+1} > \cdots$. Since $(r_1, r_2, \cdots)$ is a decreasing sequence of integers, there is some integer n such that $r_{n+1} = 0$.

The above fact can be expressed by the following equations,

$$\begin{aligned} &a - q_1b = r_1 \\ &b - q_2r_1 = r_2 \\ &r_1 - q_3r_2 = r_3 \\ &\cdots\cdots \\ &r_{n-2} - q_nr_{n-1} = r_n \\ &r_{n-1} - q_{n+1}r_n = 0 \end{aligned}$$

Put $r_n = h$. It is clear from the last equation that $h|r_{n-1})$. Since $h|r_n$ and $h|r_{n-1})$ it follows that $h|r_{n-2})$. Similarly it follows that $h|r_{n-3}, h|r_{n-4}, \cdots, h|r_2, h|r_1$. Thus $h|b$ and $h|a$.

On the other hand if $d|a$ and $d|b$ then d divides $r_1 = a - q_1b$. Again since $d|b$ and $d|r_1$, d divides $r_2 = b - q_2r_1$. Similarly it follows that d divides $r_3 = r_1 - q_3r_2$. Continuing in this way finally we observe that d divides $h = r_n = r_{n-2} - q_nr_{n-1}$. Thus $(a, b) = h$ and finally we observe that,

$$\begin{aligned} h &= r_n = r_{n-2} - q_nr_{n-1} = r_{n-2} - q_n(r_{n-3} - q_{n-1}r_{n-2}) \\ &= (-q_n)r_{n-3} + (q_nq_{n-1} + 1)r_{n-2} \\ &= (-q_n)r_{n-3} + (q_nq_{n-1} + 1)(r_{n-4} - q_{n-2}r_{n-3}) \\ &= (q_nq_{n-1} + 1)r_{n-4} + (-q_n - q_{n-2} - q_nq_{n-1}q_{n-2})r_{n-3} \\ &= \cdots = aP_n + bQ_n. \end{aligned}$$

For example, consider 741 and 1079. Then

```
741)1079(1
     741
---------
338)741(2
    676
---------
65)338(5
   325
---------
13)65(5
   65
---------
    0
```

Hence $(741, 1079) = 13$. Further

$$\begin{aligned} 13 &= 338 - 325 \\ &= 338 - 5(65) \\ &= 338 - 5\{741 - 2(338)\} \\ &= 11(338) - 5(741) \\ &= 11(1079 - 741) - 5(741) \\ &= 1079(11) + 741(-16) \\ &= a_1x_1 + a_2x_2 \end{aligned}$$

where, $x_1 = 11$ and $x_2 = -16$.

We now discuss the Lowest Common Multiple (L.C.M) of two positive integers a and b.

Definition

If m is such an integer that (i) $a|m, b|m$ and (ii) if $a|c$, $b|c$ then $m|c$, then m is said to be the L.C.M. of a and b and is written as $m = [a, b]$. It is clear that there cannot be more than one integer satisfying the above two conditions. We now prove that $[a, b](a, b) = a.b$.

Proof: Let $(a, b) = d$, $a = a_1d$ and, $b = b_1d$ and $(a_1, b_1) = 1$. It now suffices to prove that $[a, b] = a_1b_1d = \frac{a.b}{d}$. Clearly $a|ab_1 = a_1b_1\,d$ and $b|ab_1 = a_1b_1\,d$.

Secondly if $a|c$ and $b|c$ then $c = c_1$, $a = c_1a_1d$, $c = c_1b = c_1b_1d$. So $c_1a_1 = c_1b_1$. Since $(a_1, b_1) = 1$, $b_1\,|c_1$. Thus take $c_1 = kb_1$. So $c = c_1a_1d = kb_1a_1d$, i.e. $a_1b_1\,d|c$, so $a_1b_1d = [a, b]$ and $ab = [a, b].(a, b)$.

Definition

Let $a_1, a_2, \cdots a_k$ be given positive integers. Let m be such an integer that,

(i) $a_i \mid m$ for $i = 1, 2, \cdots, k$

(ii) If $a_i \mid c$ for $i = 1, 2, \cdots, k$ then $m \mid c$.

Then m is said to be the L.C.M of $a_1, a_2, \cdots a_k$ and is written as,

$$m = [a_1, a_2, \cdots a_k].$$

We will prove that if for $i = 1, 2, \cdots, k$, $a_i = p_1^{x_{i1}} p_2^{x_{i2}} \cdots p_r^{x_{ir}}$ (p_m being prime and $x_{ij} \geq 0$) and for $j = 1, 2, \cdots, r$, v_j is the greatest of all x_{ij} $(i = 1, 2, \cdots, k)$ then

$$[a_1, a_2, \cdots a_k] = p_1^{v_1} p_2^{v_2} \cdots p_r^{v_r}$$

Proof: Clearly for each i, $a_i = p_1^{v_1} p_2^{v_2} \cdots p_r^{v_r}$. If for $i = 1, 2, \cdots, k$, $a_i \mid c$ and if for $i = s, x_{i1} = v_1$, then from $a\_s \mid c$ we get $p_1^{v_1} \mid c$. Similarly, each one of $p_2^{v_2} p_3^{v_3} \cdots p_r^{v_r}$ divides c and hence $p_1^{v_1} p_2^{v_2} \cdots p_r^{v_r} \mid c$.

In this section we discuss some properties of prime numbers.

2.5.4 *Theorem* *(Euclid's Second Theorem)*

There are infinitely many prime numbers.

Proof: Let p be a prime number. It is enough to show that there is a prime number greater than p. Let,

$$q = (2.3, 5.7, 7.11 \cdots p) + 1$$

Now q is not divisible by any one of $2, 3, 5, 7, 11, \cdots, p$. Therefore, every prime factor of q is greater than p, i.e. either q is divisible by a prime number great than p or q itself is prime. This completes the proof.

It may be noted here that every prime number greater than 2 is an odd integer, so it can be expressed by $4n + 1$ or $4n + 3$.

There are infinitely many prime numbers which can be expressed in the form $4n + 3$.

Proof: Firstly, $3 = 4.0 + 3$ is a prime number. If $Q = (2^2.3.5.7.11 \cdots p) + 1$, then Q can be expressed in the form $(4n + 3)$. If none of $2, 3, \cdots$, p divide Q, so every prime factor of Q is greater than p. On the other hand Q cannot be expressed as a product of prime numbers of the form $(4n + 1)$, since $(4n_1 + 1)(4n_2 + 1) \cdots (4n_k + 1) = (4n + 1)$. Thus Q will be divisible by a prime number greater than p and expressible in the form $(4n + 3)$.

Therefore given a prime p there always exists a prime number greater than p and expressible in the form $(4n+3)$ and this proves the assertion.

Congruence

Definition

Let m be a positive integer and $m|(x-a)$. Then x is said to be congruent to $a\,modm$ and written as $x \equiv a(modm)$. Then a is said to be a residue of $xmodm$. Again if $0 \leq a \leq m-1$, then a is said to be the least residue of $xmodm$.

For a given residue, the set of all integers congruent to this $modm$ is called the *class of residues* $modm$. The difference of any two elements of this class is divisible by m.

2.5.4 ***Example:*** Let $m = 4, a = 3$. Then if $x \equiv 3\ (mod\ 4), 4|(x-3)$, i.e., $x - 3 = 4q$ or $x = 4q + 3$. By taking different integer values of q we obtain the set of values of x:

$$\{\cdots,-17,-13,-9,-5,-1,3,7,11,15,19,\cdots\}$$

We denote this set by $\overline{3}$. Similarly,

$$\overline{0} = \{\cdots,-4,0,4,8,\cdots\}$$
$$\overline{1} = \{\cdots,-7,-3,1,5,9,\cdots\}$$
$$\overline{2} = \{\cdots,-6,-2,2,6,10,\cdots\}$$

Similarly, if $m = 5$ we get

$$\overline{0} = \{\cdots,-15,-10,-5,0,5,10,\cdots\}$$
$$\overline{1} = \{\cdots,-14,-9,-4,1,6,11,\cdots\}$$
$$\overline{2} = \{\cdots,-13,-8,-3,2,7,12,\cdots\}$$
$$\overline{3} = \{\cdots,-12,-7,-2,3,8,13,\cdots\}$$
$$\overline{4} = \{\cdots,-11,-6,-1,4,9,14,\cdots\}$$

It is clear that for a given m, the m number of residue classes are as follows:

$$0^{-} = \{\cdots,-3m,-2m,-m,0,m,2m,3m,\cdots\}$$
$$1^{-} = \{\cdots,-3m+1,-2m+1,-m+1,1,m+1,2m+1,3m+1,\cdots\}$$
$$2^{-} = \{\cdots,-3m+2,-2m+2,-m+2,2,m+2,2m+2,3m+2,\cdots\}$$
$$\cdots\cdots\cdots$$
$$r^{-} = \{\cdots,-3m+r,-2m+r,-m+r,r,m+r,2m+r,3m+r,\cdots\}$$
$$(m-1)^{-} = \{\cdots,-2m-1,-m-1,-1,m-1,2m-1,3m-1,\cdots\}$$

The set consisting of $\bar{0}, \bar{1}, \bar{2}, \cdots, (m-1)^{-}$ or the set formed by choosing one representative from each of the above sets is called a complete system of incongruent residues.

Some properties of congruence: It is clear that for a given m,

(a) $a \equiv a$ (reflexivity),
(b) $a \equiv b \Rightarrow b \equiv a$ (symmetry),
(c) $a \equiv b$ and $b \equiv c \Rightarrow a \equiv c$ (transitivity),
(d) $a \equiv b$ and k is any integer $\Rightarrow ka = kb$,
(e) $a \equiv b$ and $a' \equiv b' \Rightarrow a + a' \equiv b + b'$ and $aa' \equiv bb'$,

Further for any positive integer n, $a^n \equiv b^n$.

Note that for a given m, the relation. "Congruence mod m" is an equivalence relation and hence defines a partition in the set of all integers.

Examples of Proof by Induction

Prove the following statements by induction on n:

1. $\sum_{i=1}^{n} i = \frac{1}{2} n(n+1), \ (n \geq 1)$
2. $\sum_{i=1}^{n} i^2 = \frac{1}{6} n(n+1)(2n+1), (n \geq 1)$
3. $\sum_{i=1}^{n} i^3 = \left[\frac{1}{2} n(n+1)\right]^2, \ (n \geq 1)$
4. $(A_1 \cup A_2 \cdots \cup A_n)' = A_1' \cap A_2' \cdots \cap A_n', \ (n \geq 1)$
5. $(A_1 \cap A_2 \cdots \cap A_n)' = A_1' \cup A_2' \cdots \cup A_n', \ (n \geq 1)$
6. For $a, \ b \in R$ and $n \geq 1$,

   $(a+b)^n = a^n + {}^nC_1 \, a^{n-1}b + {}^nC_2 \, a^{n-2}b^2 + \cdots + b^n.$

We prove only 1, 4 and 6; others are left to the reader as exercises.

***Proof of* (1):** For $n = 1$, the statement is true, because in this case,

$$\sum_{i=1}^{1} i = \frac{1}{2} 1(1+1) = \frac{1}{2}.1.2 = 1.$$

The induction hypothesis is that for all $k \geq 1$,

$$\sum_{i=1}^{k} i = \frac{1}{2} k(k+1)$$

Now,

$$\sum_{i=1}^{k+1} i = \left(\sum_{i=1}^{k} i\right) + (k+1)$$

$$\left(\frac{1}{2}k(k+1)\right) + (k+1) = \frac{1}{2}(k+1)(k+2)$$

which proves the statement for $n = k+1$. This completes the proof.

***Proof of* (4):** For $n=1$, the statement is trivially true. The induction hypothesis is that for all $k \geq 1$,

$$(A_1 \cup A_2 \cdots \cup A_k)' = A_1' \cap A_2' \cdots \cap A_k'$$

Using De-Morgan's laws we have

$$\begin{aligned}(A_1 \cup A_2 \cdots \cup A_{k+1})' &= \left[(A_1 \cup A_2 \cdots \cup A_k) \cup A_{k+1}\right]' \\ &= (A_1 \cup A_2 \cdots \cup A_k)' \cap A_{k+1}' \\ &= A_1' \cap A_2' \cdots \cap A_k' \cap A_{k+1}'\end{aligned}$$

which proves the statement for $n = k+1$ and this completes the proof.

***Proof of* (6):** For $n=1$, the statement is clearly true. The induction hypothesis is that for all $k \geq 1$,

$$(a+b)^k = a^k + {}^kC_1 a^{k-1} b + \cdots + b^k$$

Now

$$\begin{aligned}(a+b)^{k+1} &= (a+b)^k (a+b) = \left(a^k + {}^kC_1 a^{k-1} b + \cdots + b^k\right)(a+b) \\ &= a^{k+1} + \left({}^kC_1 + 1\right) a^k b + \cdots + b^{k+1} = a^{k+1} + {}^{k+1}C_1\, a^k b + \cdots + b^{k+1}\end{aligned}$$

which proves the statement for $n = k+1$ and this completes the proof.

2.6 NUMERATION SYSTEMS WITH DIFFERENT BASES

The base of a numeration system is equal to the number of basic symbols used to describe the system. The numeration system with symbols 0, 1, 2, 3, 4, 5, 6, 7, 8, 9 is called the decimal or base ten system. The numbers in decimal system can be expressed in the following manner:

$$273 = 2 \times 10^2 + 7 \times 10 + 3 \times 1$$

$$10925 = 1 \times 10^4 + 0 \times 10^3 + 9 \times 10^2 + 2 \times 10 + 5 \times 1$$

Similarly, only the digits 0, 1, 2, 3, 4 are used in the base five system; the digits 0, 1, 2 are used in the ternary (base three) system; the digits 0, 1 are used in the binary (base two) system. The latter system is used in the computer for numeric computation. We write $(101101)_2$ to denote the number 101101 in binary system and use similar notation for numbers with different bases. Only numbers in decimal system are written without the mention of their bases.

2.6.1 *Examples*

1. $(101101)_2 = 1 \times 2^5 + 0 \times 2^4 + 1 \times 2^3 + 1 \times 2^2 + 0 \times 2 + 1$
   $= 32 + 0 + 8 + 4 + 1 = (45)_{10}$
2. $(43)_5 = 4 \times 5 + 3 = 20 + 3 = (23)_{10}$
3. $(432.2)_8 = 4 \times 8^2 + 3 \times 8 + 2 + 2 \times 8^{-1} = (282.25)_{10}$
4. $(1101.01)_2 = 1 \times 2^3 + 1 \times 2^2 + 0 \times 2 + 1 + 0 \times 2^{-1} + 1 \times 2^{-2} = (13.25)_{10}$

In a system with base, a positive number N represents the polynomial

$$(N)_b = a_{q-1}\ b^{q-1} + \cdots + a_0 b^0 + \cdots + a_{-p} b^{-p} = \sum_{i=-p}^{q-1} a_i b^i$$

where the base b is an integer >1 and each a_i is an integer in the range $0 \le a_i \le b-1$. The sequence of digits $a_{q-1}\, a_{q-2} \cdots a_0$ constitutes the integral part of N, while the sequence $a_{-1}\, a_{-2} \cdots a_{-p}$ constitutes the fractional part of N. Thus, p and q designate the number of digits in the fractional part and integral part, respectively. The digit a_{-p} is called the least significant digit while a_{q-1} is called the most significant digit. When $b_1 > b_2$, it is more convenient to use base b_{-1} arithmetic. The conversion is done by considering separately the integral and fractional parts of N. Let $(N)_{b_1}$ be an integer whose value in base b_1 is given by,

$$(N)_{b_1} = a_{q-1} b_2^{q-1} + a_{q-2} b_2^{q-2} + \cdots + a_1 b_2^1 + a_0 b_2^0$$

To find the values of the a_i's, let us divide (base b_1) the above polynomial by b_2.

$$\frac{(N)_{b_1}}{b_2} = a_{q-1} b_2^{q-2} + a_{q-2} b_2^{q-3} + \cdots + a_1 + \frac{a_0}{b_2}$$

Thus, the least significant digit of $(N)_{b_1} = a_0$ is equal to the first remainder. The next significant digit a_1 is obtained by dividing the quotient Q_0 by b_2 i.e.,

$$\frac{(Q_0)}{b_2} b_1 = a_{q-1} b_2^{q-2} + a_{q-2} b_2^{q-3} + \cdots + \frac{a_1}{b_2}$$

The remaining a_i's are evaluated by repeated divisors of the quotients until Q_{q-1} is equal to zero.

2.6.7 *Examples*

1. Convert $(0.3125)_{10}$ to base 8

   $0.3125 \times 8 = 2.5000$ hence $a_{-1} = 2$,

   $0.5000 \times 8 = 4.0000$ hence $a_{-2} = 4$

   Thus $(0.3125)_{10} = (0.24)_8$.

2. $(0.375)_{10} = (0.011)_2$

   $0.375 \times 2 = 0.750$ hence $a_{-1} = 0$,

   $0.750 \times 2 = 1.500$ hence $a_{-2} = 1$,

   $0.500 \times 2 = 1.000$ hence $a_{-3} = 1$

   Thus $(0.375)_{10} = (0.011)_2$.

3. $(123.4)_8 = (001010011.100)_2$

4. $(1010110.0101)_2 = (126.24)_8$.

Binary Arithmetic

The following table gives elementary binary operations:

Binary Digits		*Sum*	*Carry*	*Difference*	*Borrow*	*Product*
a	*b*	*a + b*		*a – b*		*a . b*
0	0	0	0	0	0	0
0	1	1	0	1	1	0
1	0	1	0	1	0	0
1	1	0	1	0	0	1

2.6.8 *Examples*

1. 1111 = carries of 1

$$\begin{array}{r} 1111.01 = (15.25)_{10} \\ +\,0111.10 = (7.50)_{10} \\ \hline 10110.11 = (22.75)_{10} \end{array}$$

2. 1 = borrows of 1

$$\begin{array}{r} 10010.11 = (18.75)_{10} \\ -\,01100.10 = (12.50)_{10} \\ \hline 00110.01 = (6.25)_{10} \end{array}$$

3.

$$\begin{array}{r} 11001.1 = (25.5)_{10} \\ \times \quad 110.1 = (6.5)_{10} \\ \hline \end{array}$$

$$\begin{array}{l} \quad\;\; 110011 \\ \quad\; 000000 \\ \;\; 11001 \\ \underline{110011} \\ 10100101.11 = (1655.75)_{10} \end{array}$$

4. Divide $(1000100110)_2$ by $(11001)_2$.

10110 = quotient

$$\begin{array}{r} 11001\overline{)1000100110} \\ \underline{11001} \\ 00100101 \\ \underline{11001} \\ 0011001 \\ \underline{11001} \\ 00000 = \text{remainder} \end{array}$$

EXERCISES

1. Let $f: X \to Y$ be an arbitrary mapping. Define a relation in X as follows: $x_1 \sim x_2$ means that $f(x_1) = f(x_2)$. Show that this is an equivalence relation and describe the equivalence sets.
2. In the set R of all real numbers, let $x \sim y$ mean that $x - y$ is an integer. Show that this is an equivalence relation and describe the equivalence sets.
3. Let I be the set of all integers, and let m be affixed positive integer. Two integers a and b are said to be congruent modulo m symbolized by $a \equiv b\,(\text{modm})$ if $a–b$ is exactly divisible by m, i.e., if $a - b$ is an integral multiple of m. Show that this is an equivalence relations, describe the equivalence sets.
4. Decide which ones of the four properties are reflexivity, antisymmetry, symmetry, and transitivity are true for each of the following relations in the set of positive integers: $m \leq n, m < n, m$ divides n, $m = n$.
5. Let X be a non-empty set. A relation $\sim$ in X is called circular if $x \sim y$ and $y \sim z \Rightarrow z \sim x$, and triangular if $x \sim y$ and $x \sim z \Rightarrow y \sim z$. Prove that a relation in X is an equivalence relation $\Leftrightarrow$ it is reflexive and circular $\Leftrightarrow$ it is reflexive and triangular.
6. Give an example of a relation which is (a) reflexive but not symmetric or transitive; (b) symmetric but not reflexive or transitive, (c) transitive but not reflexive or symmetric; (d) reflexive and transitive but not symmetric; (f) symmetric and transitive but not reflexive.
7. Let X be a non-empty set, and let f and g be one-to-one mappings of X onto itself. Show that fg is also a one-to-one mapping of X onto itself and that $(fg)^{-1} = g^{-1}f^{-1}$.
8. Let f, g and h be any three mappings of a non-empty set X into itself, and show that multiplication of mappings is associative in the sense that $f(gh) = (fg)h$.
9. Let X and Y be non-empty sets and f a mapping of X into Y. Show the following:

   (a) If f is one to one $\Leftrightarrow$ there exists a mapping g of Y into X such that, $gf = idx$;

   (b) f is onto $\Leftrightarrow$ there exists a mapping h of Y into X such that $fh = idy$.
10. Let X be a non-empty set and f a mapping of X into itself. Show that f is one-to-one onto $\Leftrightarrow$ there exists a mapping g of X into itself such that $fg = gf = idx$.
11. Prove the following results by mathematical induction.

    (a) $\sum_{i=1}^{n} i = \frac{1}{2}n(n+1)(n \geq 1)$.

(b) $\sum_{i=1}^{n} i^2 = \frac{1}{6}n(n+1)(2n+1)$ $(n \geq 1)$.

(c) $\sum_{i=1}^{n} i^3 = \left[\frac{1}{2}n(n+1)\right]^2$ $(n \geq 1)$.

(d) $1.2 + 2.3 + \cdots + n(n+1) = \frac{n(n+1)(n+2)}{3}$.

(e) $1^2 + 3^2 + 5^2 + \cdots + (2n-1)^2 = \frac{n(2n+1)(2n-1)}{3}$.

(f) $2 + 5 + 8 + \cdots + (3n-1) = \frac{3(3n+1)}{2}$.

(g) $1 + 5 + 9 + \cdots + (4n-3) = n(2n-1)$.

(h) $2 + 2^2 + \cdots + 2^n = 2^{n+1} - 2$.

(i) $(a+b)^n = a^n + \binom{n}{1}a^{n-1}b + \binom{n}{2}a^{n-2}b^2 + \cdots \ldots . + b^n$.

(j) $2^n > n^3$for $n \geq 10$.

(k) $1 + 2^n < 3^n$for $n \geq 2$.

(l) $2^n > n$ for $n > 2$.

(m) Any positive integer $n > 2$ is either a prime or a product of primes.

(n) $n^3 + 2n$ is divisible by 3 for $n \geq 2$.

(o) $n^4 - 4n^2$ is divisible by 3 for $n \geq 2$.

12. Convert the following

(a) $(1431)_8$ to base 10.

(b) 11001010.0101 to base 10.

(c) 11001101.0101 to base 8 and 4.

(d) $(1984)_{10}$ to base 8.

(f) $(1776)_{10}$ to base 6.

(g) $(53.1575)_{10}$ to base 2.

(h) $(3.1415)_{10}$ to base 8 and base 2.

13. Given the binary number $a = 1010.1$, $b = 101.01$, $c = 1001.1$, perform the following binary operations:

(a) $a + c$

(b) $a - b$

(c) $a.c$

(d) $\frac{a}{b}$

14. Each of the following arithmetic operations is correct in at least the number system. Determine the possible bases of the numbers in each operation:
    (a) $1234 + 5432 = 6666$
    (b) $\frac{41}{3} = 13$
    (c) $\frac{33}{3} = 11$
    (d) $23 + 44 + 14 + 32 = 233$
    (e) $\frac{30}{20} = 12.1$
    (f) $\sqrt{41} = 5$.
15. If for $i \neq j$, $(a_i, a_j) = 1$ then prove that
    $[a_1, a_2, \cdots, a_k] = a_1 a_2 \cdots a_k$.
16. Prove that,
    $[a_1, a_2, \cdots, a_k] = [[a_1 a_2 \cdots, a_{k-1}], a_k]$
17. If $(a, b) = 1$ and c/a then show that $(c, b) = 1$.
18. Prove that,
    $(ma, mb) = m(a, b)$
19. If $(a, b) = 1$ then show that,
    $(ac, b) = (c, b)$

CHAPTER 3

Combinatorics

The numeric computation always deals with fundamental principles of counting. There are some situations where these rules are not applied directly and some indirect techniques are used to overcome these types of counting difficulties. Combinatorics is a branch of mathematics that deals with counting techniques and their algorithms. In this chapter we briefly discuss the basic concepts of combinatorics. The topics covered are: Permutations and combinatorics, Pigeonhole Principle, Numeric Functions, Generating Functions and Recurrence Relations.

3.1 PERMUTATIONS AND COMBINATIONS

In many counting problems we want to determine the number of different ways of arranging or ordering a set of distinct objects.

If S is a set, then a permutation of S is a bijection of S onto itself, that is, a permutation of S is a mapping $Q : S \to S$, which is both injective and surjective. For example, $id : S \to S$ is a permutation of S. In other words, given a set S, an arrangement of the elements of S in a definite order without repetitions is called a permutation of the set S.

3.1.1 ***Example:*** There are two different permutations of the set $\{a, b\}$. They are $a \to a,\ b \to b$ and $a \to b,\ b \to a$. Since this notation is cumbersome, we write out permutations as,

$$\begin{pmatrix} a & b \\ a & b \end{pmatrix}, \begin{pmatrix} a & b \\ b & a \end{pmatrix}$$

3.1.2 ***Example:*** There are six different permutations of the set $\{1, 2, 3\}$. They are

$$\begin{pmatrix} 1 & 2 & 3 \\ 1 & 2 & 3 \end{pmatrix}, \begin{pmatrix} 1 & 2 & 3 \\ 2 & 1 & 3 \end{pmatrix}, \begin{pmatrix} 1 & 2 & 3 \\ 3 & 1 & 2 \end{pmatrix}$$

$$\begin{pmatrix} 1 & 2 & 3 \\ 1 & 3 & 2 \end{pmatrix}, \begin{pmatrix} 1 & 2 & 3 \\ 2 & 3 & 1 \end{pmatrix}, \begin{pmatrix} 1 & 2 & 3 \\ 3 & 2 & 1 \end{pmatrix}$$

3.1.3 ***Example:*** A = $\{a_1, a_2, \ldots, a_n\}$

be a set of n elements. Then a permutation of A is a mapping,

$$\theta = \begin{pmatrix} a_1 & a_2 & \cdots & a_n \\ a_{i_1} & a_{i_2} & \cdots & a_{i_n} \end{pmatrix}$$

where a_{ik} is the image of a_k under θ and $a_{i1}, a_{i2}, \ldots, a_{in}$ must be distinct elements of A, because θ is injective.

a_{i_1} can be chosen in any one of the n ways; then since θ is injective, a_{i_2} can be selected in any one of $(n-1)$ ways, then a_{i3} in any of $(n-2)$ ways etc., giving $n(n-1)(n-2)\ldots 3.\ 2.\ 1$ different permutations of A.

Let $S(n)$ denote the set of all permutations, then $S(n)$ has $n!$ elements where $n!$ (read n factorial) denotes the product of the integers from 1 to n. Thus $S(1)$ has one element, $S(2)$ has two elements and $S(3)$ has six elements and so on.

Once the value of $n!$ is known, $(n+1)!$ can be obtained as follows,

$$(n+1)! = (n+1)n!$$

Since $n!$ is defined to be the product of the integers from 1 up to n, the symbol 0! is not yet defined. We agree to give 0! the value 1 and with this definition the above formula also holds when $n = 0$.

In forming a permutation of a set, all the elements of the set are used. In some problems, however, we want to consider arrangements where not all the elements are used. If we want a permutation of r elements selected from a set of n elements, then this is called a permutation of the n objects taken r at a time.

We shall now derive a formula for the number of permutations of n objects taken r at a time. Since we have n objects, the first object in the permutation can be chosen in n ways; then since repetitions are not allowed, the second object can be chosen in $(n-1)$ ways, then the next object can be chosen in $(n-2)$ ways and so on. Since each permutation contains r objects, it follows from the multiplication principle that there are,

$$\underbrace{n(n-1)(n-2)\ldots}_{\text{rfactors}}$$

different permutations. We write n_{P_r} to denote the number of permutations of n distinct objects taken r at a time and observe that the rth or last factor in $n(n-1)(n-2)\ldots$ is $n-(r-1) = n-r+1$. Therefore we have,

$$n_{P_r} = n(n-1)(n-2)\ldots(n-r+1)$$

$$= \frac{n(n-1)(n-2)\ldots(n-r+1)(n-r)!}{(n-r)!} = \frac{n!}{(n-r)!}$$

3.1.4 ***Example:*** Compute 8_{P_3}

$$8_{P_3} = 8.7.6 = 336$$

or $$8_{P_3} = \frac{8!}{5!}$$

$$= \frac{8.7.6.5!}{5!} = 8.7.6 = 336.$$

In many counting problems we want to determine the number of different subsets with r objects that can be selected from a set with n objects.

To illustrate why one might be interested in counting subsets, suppose we are interested in finding the number of different five-card hands that can be dealt from a standard deck of 52 cards. A hand of five cards can be viewed as a set of five elements selected from a set of 52 elements. Thus, the number of different five-card hands that can be dealt is the same as the number of different subsets with five elements that can be selected from a set with 52 elements.

3.1.5 ***Example:*** There are four subsets with three objects that can be selected from the set,

$$\{1, 2, 3, 4\}.$$

They are

$$\{1, 2, 3\}, \{1, 2, 4\}, \{1, 3, 4\}, \{2, 3, 4\}$$

But since $4_{P_3} = 4!$, there are 24 permutations of $\{1, 2, 3, 4\}$ taken three at a time. There are many more permutations with three objects than subsets with three elements, since

$$\begin{pmatrix} 1 & 2 & 3 & 4 \\ 3 & 2 & 4 & 1 \end{pmatrix} \text{ and } \begin{pmatrix} 1 & 2 & 3 & 4 \\ 4 & 2 & 1 & 3 \end{pmatrix}$$

are different permutations of $\{1, 2, 3, 4\}$ taken three at a time while the sets $\{3, 4, 1\}$ and $\{4, 1, 3\}$ are not different.

A subset of r elements selected from a set of n elements is called a combination of n objects taken r at a time. We now develop a formula for the number of different combinations of n objects taken r at a time and we denote this quantity by the symbol n_{Cr}. We often use the symbol $\binom{n}{r}$ for n_{Cr}.

To obtain a formula for n_{Cr}, we need only observe that all permutations of n objects taken r at a time can be obtained by

1. Listing all combinations of n objects taken r at a time.
2. Taking each combination and listing all the permutations of its elements.

3.1.6 ***Example:*** The combinations of:

$\{1, 2, 3, 4\}$

taken three at a time are,

$\{1, 2, 3\}, \{1, 2, 4\}, \{1, 3, 4\}, \{2, 3, 4\}$

Now each of these combinations has six permutations of its elements, and there are 24 permutations of $\{1, 2, 3, 4\}$ taken three at a time.

Suppose we are given a set with n elements, n_{C_r} denotes the number of combinations taken r at a time. Since each of these combinations has r elements, the elements can be permuted in $r!$ ways. Hence by the multiplication principle, the number of forming a combination and permuting its elements is,

$n_{C_r} . r!$

since this process gives all permutations of n elements taken r at a time, we obtain,

$$n_{P_r} = n_{C_r} .r!$$

or,

$$n_{C_r} = \frac{n_{P_r}}{r!} = \frac{n!}{r!(n-r)!}.$$

Thus the number of combinations of n elements taken r at a time is given by,

$$n_{C_r} = \frac{n!}{r!(n-r)!}$$

3.1.7 ***Example:*** How many different five-card hands can be dealt from a deck of 52 cards?

Solution: The number of five-card hands is the number of combinations of 52 objects taken five at a time. Therefore,

$$52_{C_5} = \frac{52!}{5!47!} = \frac{52.51.50.49.48}{5.4.3.2.1} = 2{,}598{,}960$$

3.1.8 ***Example:*** Show that the number of ways to select none, one, two,... or n objects from n objects is 2^n.

Observe that

$$(*)\ A(z) = (1+z)^n = n_{C_0} + n_{C_1} z + n_{C_2} z^2 + \ldots + n_{C_n} z^n$$

Putting $z = 1$ in the above identity we get,

$$n_{C_0} + n_{C_1} + n_{C_2} + \ldots + n_{C_n} = (1+1)^n = 2^n.$$

But, $n_{C_0}+n_{C_1}+n_{C_2}+\ldots+n_{C_n}$ is the number of ways to select none, one, two,... or n objects from n objects and hence, the result.

Note that an alternative way of putting the above question is that the number of subsets of a finite set with n elements is 2^n.

3.1.9 ***Example:*** Show that the number of ways of selecting an even number of objects from n objects is equal to the number of ways of selecting an odd number of objects.

Putting $z = -1$ in (*) we get,

$$n_{C_0}-n_{C_1}+n_{C_2}+\ldots+(-1)^r n_{Cr}+\ldots+(-1)^n n_{C_n}=0$$

Or

$$n_{C_0}+n_{C_2}+n_{C_4}\ldots=n_{C_1}+n_{C_3}+n_{C_5}+\ldots.$$

3.1.10 ***Example:*** Show that,

$$n_{C_r}=n-1_{C_r}+n-1_{C_{r-1}}$$

We have

$$\begin{aligned} n_{C_r} &= \frac{n!}{r!(n-r)!} \\ &= \frac{n!}{r!(n-r)!}\left(\frac{n-r}{n}+\frac{r}{n}\right) \\ &= \frac{(n-1)!}{r!(n-r-1)!}+\frac{(n-1)!}{(r-1)(n-r)!} \\ &= n-1_{C_r}+n-1_{C_{r-1}}. \end{aligned}$$

We can also prove this by using combinatorial arguments as follows:

If we are to select r objects from n objects, there are $n-1_{C_r}$ ways to select r objects so that a particular object is always excluded; in $n-1_{C_{r-1}}$ ways so that this particular object is always included. Hence the result follows.

3.1.11 ***Example:*** Let a_r denote the number of ways in which r objects can be selected from n objects with unlimited repeation. Since each object can be selected as many times as we wish,

a_r = coefficient of z^r in $(1 + z + z^2 + \ldots)^n$.

Observe that the contribution from each of the factors will correspond to the number of times one of the objects is selected. Since

$$(1+z+z^2+\ldots)^n=\left(\frac{1}{1-z}\right)^n=(1-z)^n$$

We obtain

$$a_r = (-1)^r \frac{(-n)(-n-1)...(n-r-1)}{r!} = \frac{(n+r-1)!}{r!(n-1)!} = n+(r-1_{C_r}).$$

3.2 PIGEONHOLE PRINCIPLE

We shall now describe an important principle, called Pigeonhole Principle. Expressed informally, it states that if there are 'many' pigeons and 'a few' pigeonhole, then there must be some pigeonhole occupied by two or more pigeons. To be more precise, let D and R be finite sets and let $o(D) > o(R)$. Then for any function f from D to R, there exist d_1, $d_2 \in D$ such that $f(d_1) = f(d_2)$.

3.2.1 *Example*

1. If eight people are selected in any manner whatsoever, birthday of at least two of them will fall on the same day of the week. In this case each day of the week corresponds to pigeonhole.
2. If thirteen people are selected, there are at least two of them who were born in the same month. In this case 13 people are the pigeons and the 12 months are the pigeonholes.
3. Among 32 people born during the month of January, at least two of them will have the same date as their birth day.

We now discuss the Extended Pigeonhole Principle.

The Extended Pigeonhole Principle: If n pigeons are assigned to m pigeonholes, then one of the pigeonholes must contain at least $\left[\frac{n-1}{m}\right]+1$ pigeons.

Proof: If each pigeonhole contains not more than $\left[\frac{n-1}{m}\right]$ pigeons, then there are at the most,

$$m\left[\frac{n-1}{m}\right] \leq \frac{m(n-1)}{m} = n-1$$

pigeons in all. This contradicts the assumption and hence, the proof is complete.

Numeric Function

In this section we shall discuss about numeric functions. Such functions are very much useful to computer scientists because we encounter them very often in digital computation.

Let N and R respectively denote the sets of all natural numbers and real numbers. A mapping (function) from N to R is called a numeric function. Some people also call it a discrete numeric function. In fact a numeric function is a sequence of real numbers. We also include zero to the domain of a numeric function. If $a_0, a_1, \ldots, a_r, \ldots$ denote the values of the function at 0, 1, …, r, … respectively, then we express the function by

$$(a_0, a_1, a_2, \ldots, a_r, \ldots)$$

and we denote it by a.

The forward difference of a numeric function a is a numeric function denoted by Δa. It is defined as the function whose value at r is $a_{r+1} - a_r$ for $r = 0, 1, 2, \ldots$, i.e.,

$$(\Delta a)_r = a_{r+1} - a_r, \quad r = 0, 1, 2, \ldots$$

The backward difference of a is denoted by ∇a and it is defined by

$$(\nabla a)_r = a_r - a_{r-1}, \quad r = 0, 1, 2, \ldots$$

Let a and b be two numeric functions. The convolution of a and b, denoted by $a*b$ is a numeric function C where,

$$c_r = a_0 b_r + a_1 b_{r-1} + a_2 b_{r-2} + \ldots + a_r b_0$$

$$= \sum_{i=0}^{r} a_i b_{r-i}.$$

Remark It should be noted that n_{C_r}, the combination of n things taking r at a time, is a special case and hence is an example of a numeric function having a finite number of values.

To observe this, let n be a fixed positive integer. Put

$$a_r = n_{C_r} \text{ for } r = 0, 1, 2, \ldots, n.$$

Then

$$(a_0, a_1, a_2, \ldots a_r, \ldots, a_n) = (n_{C_0}, n_{C_1}, n_{C_2}, \ldots, n_{C_r}, \ldots, n_{C_n})$$

is a finite numeric function, i.e., a numeric function having values in a finite subset of the real line R.

We now discuss an important concept called the *asymptotic behaviour of* numeric functions. In many practical applications it becomes necessary to compare the behaviour of two numeric functions in the sense that which one grows faster than the other or whether the ratio of the two becomes bounded. Therefore, we introduce a formal definition which is as follows:

If a and b are two numeric functions, then we say that 'a' asymptotically dominates 'b' or 'b' is asymptotically dominated by 'a' if there exist constants $k > 0$ and $m > 0$ such that,

$b_r \leq ma_r$ for $r \geq k$.

The set of all numeric functions that are asymptotically dominated by a is denoted by $O(a)$ and $O(a)$ is called 'the order a' or 'big-oh a'. Thus if b is asymptotically dominated by a, then b belongs to $O(a)$.

We now discuss some properties of $O(a)$ through examples and these properties are useful in some applications.

3.2.2 *Example*

1. If $a = a_0 + a_1 r + a_2 r^2 + \ldots + a_n r^n$, then a is $O(r^n)$.

   Observe that,

   $$\begin{aligned} &|a_0 + a_1 r + a_2 r^2 + \ldots + a_n r^n| \\ &\quad \leq |a_0| + |a_1| r + |a_2| r^2 + \ldots + |a_n| r^n \\ &\quad \leq (|a_0| + |a_1| + |a_2| + \ldots + |a_n|) r^n \text{ (for } r \geq 1) \end{aligned}$$

   Now if $k = 1$ and $m = |a_0| + |a_1| + |a_2| + \ldots + |a_n|$ then

   $$|a_0 + a_1 r + a_2 r^2 + \ldots + a_n r^n| \leq m r^n.$$

   Hence a is $O(r^n)$.

2. If $a = r + \frac{1}{2} r^2 + \frac{1}{3} r^3$, observe that

   $$a = O(r^3),$$

   $$a = \frac{1}{3} r^3 + O(r^2),$$

   $$a = \frac{1}{3} r^3 + r^2 + O(r).$$

   We multiply numeric functions in the usual way (coordinate wise). If

   $$a = r + O\left(\frac{1}{r}\right),$$

   $$b = +O\left(\frac{1}{\sqrt{r}}\right),$$

   then $ab = +O\left(\sqrt{r}\right)$.

We have

$$ab = \left(r + \mathrm{O}\left(\frac{1}{r}\right)\right)\left(\sqrt{r} + \mathrm{O}\left(\frac{1}{\sqrt{r}}\right)\right)$$

$$= r^{3/2} + r\mathrm{O}\left(\frac{1}{\sqrt{r}}\right) + \sqrt{r}\mathrm{O}\left(\frac{1}{r}\right) + \mathrm{O}\left(\frac{1}{r}\right)\mathrm{O}\left(\frac{1}{\sqrt{r}}\right)$$

$$= r^{3/2} + \mathrm{O}\left(\sqrt{r}\right) + \mathrm{O}\left(\frac{1}{\sqrt{r}}\right) + \mathrm{O}\left(\frac{1}{r^{3/2}}\right)$$

$$= r^{3/2} + \mathrm{O}\left(\sqrt{r}\right).$$

Generating Functions

We have been expressing the numeric function as a sequence by listing all its values. We shall now discuss an alternative (and equivalent) way of representing the numeric functions, which is called the 'generating function'. To express it formally, let,

$$a = (a_0, a_1, a_2, \ldots, a_r, \ldots)$$

be a numeric function. Define an infinite series as follows:

$$a_0 + a_1 z + a_2 z^2 + \ldots + a_r z^r + \ldots.$$

We write this as $A(z)$ and represent it as,

$$\sum_{r=0}^{\infty} a_r z^r.$$

Such a series is called a power series. This series is called the generating function of the numeric function 'a'. Observe that given a numeric function 'a' we have the infinite series $\sum_{r=0}^{\infty} a_r z^r$ and conversely, given an infinite series of the form $a = \sum a_r z^r$, we have a numeric function $a = (a_0, a_1, a_2, \ldots, a_r, \ldots)$. Hence, a numeric function can be represented by a 'generating function' in an equivalent manner.

In many situations alternative representations are very much convenient and useful. Expressing a decimal number in binary is an alternative arrangement and this helps us to do the arithmetic operation by a computer. Taking logarithm is an alternative representation of numbers and this helps us to do computations in easier way because multiplication reduces to addition.

3.2.3 *Examples*

1. Let a be any numeric function and b the numeric function (1, 1, 1, ...). Let us obtain the convolution of a and b.

   Now if,

   $$c = a * b,$$

   then,

   $$c_r = \sum_{i=0}^{r} a_i b_{r-i} = \sum_{i=0}^{r} a_i.$$

   The generating function $c(z)$ of c is

   $$c(z) = \frac{1}{z} A(z).$$

   If $A(z) = \frac{1}{1-z}$, then the generating function of the numeric function (1, 2, 3, ...) is given by $\frac{1}{(1-z)^2}$.

2. Let $a_r = 3^r$, $r \geq 0$ and $b_r = 2^r$, $r \geq 0$.

   Now $A(z) = \frac{1}{1-3z}$ and $B(z) = \frac{1}{1-2z}$.

   Let $c = a * b$.

   Then $C(z) = A(z)B(z) = \left(\frac{1}{1-3z}\right)\left(\frac{1}{1-2z}\right)$

   $$= \frac{3}{1-3z} - \frac{2}{1-2z}$$

   $$C_r = 3(3^r) - 2(2^r) = 3^{r+1} - 2^{r+1}.$$

3.3 RECURRENCE RELATIONS

Before we define Recurrence Relation precisely we first discuss some examples. Suppose one has a friend who has three sons. If he is asked about the age of his eldest son, he may reply in two different ways. One way replying is the straightforward answer that the age of his eldest son is 20. Another way of reply is that his eldest son is 5 years older than his middle son. Then one will enquire about the age of middle son. He may then reply that it is 5 years more than his youngest son. When asked about the age of youngest son, he replies that it is 10. So knowing the age of his youngest son, one not only knows the age of his eldest son, but one knows the age of all his sons.

Now we come to the precise mathematical description.

Suppose we want to know completely a numeric function,

$$(a_0, a_1, a_2, ...).$$

If all the values a_0, a_1, a_2, ... are given then we know it completely. Instead of giving all the values, one may give us one or two values and a relation such that we can carryout a step-by-step computation to determine a_r from a_{r-1}, a_{r-2}, ...; to determine a_{r+1} from a_r, a_{r-1}, ...; and so on. The computation can be initiated because the value of the function at one or more points is given. Such a relation is called a recurrence relation and the given values of the function are called boundary conditions. Observe that a numeric function can be expressed as a recurrence relation together with an appropriate set of boundary conditions and this is yet another way of expressing a numeric function. We now describe some specific examples.

3.3.1 *Examples*

1. Consider the numeric function

$$a = (2^0, 2^1, 2^2, 2^3, ..., 2^r, ...).$$

This can be expressed as

$$a_r = 2^r, \; r = 0, 1, 2, ...$$

This can also be expressed as the generating function,

$$A(z) = \frac{1}{1-2z}.$$

Observe that a_r is 2 times a_{r-1}, a_{r-1} is 2 times a_{r-2}. Hence, we can calculate a_r if we know a_{r-1} and a_{r-1} if we know a_{r-2} and so on. Finally knowing a_0 we can find a_r for all $r = 1, 2, ...$ Thus, above numeric function can be expressed as a recurrence relation,

$$a_r = 2a_{r-1}$$

with the boundary condition $a_0 = 1$.

2. Similarly if,

$$a = (3^0, 3^1, 3^2, ..., 3^r, ...),$$

then $a_r = 3^r$.

The generating function for this is given by,

$$A(z) = \frac{1}{1-3z}$$

and the recurrence relation is given by,

$$a_r = 3a_{r-1}$$

with the boundary condition $a_0 = 1$.

3. *Fibonacci Sequence* is another important example of a recurrence relation. We discuss a sequence,

$$a = (a_0, a_1, a_2, \ldots, a_r, \ldots).$$

of numbers called the Fibonacci sequence. The first two terms of the sequence are 1 each, and each subsequent term is the sum of its two immediate predecessors. Hence, the sequence is as follows:

1, 1, 2, 3, 5, 8, 13, 21, 34....

In this case,

$$a_r = \frac{1}{5}\left(\frac{1+5}{2}\right)^{r+1} - \frac{1}{5}\left(\frac{1-5}{2}\right)^{r+1}.$$

The generating function is given by,

$$A(z) = \frac{1}{1-z-z^2}$$

and the recurrence relation is,

$$a_r = a_{r-1} + a_{r-2}$$

with the boundary conditions $a_0 = 1$, $a_1 = 1$.

EXERCISES

1. Let a be a numeric function such that,

$$a_r = \begin{cases} 2, 0 \le r \le 3 \\ 2^{-r} + 5, r \ge 4. \end{cases}$$

Determine Δa and ∇a.

2. Let a be a numeric function. Let $\Delta^{-1}a$ denote the accumulated sum of a. Are the two numeric functions $\Delta(\Delta a)$ and $\Delta^{-1}(\Delta a)$ the same?

3. Simplify the expression,

$$\left[\ln r + O\left(\frac{1}{n}\right)\right]\left[n + O\left(\sqrt{n}\right)\right].$$

4. Let a ln r. Show that for $e > 0$, a is $O(r^e)$.
5. Determine the generating function of the numeric function a_r, where

$$a_r = \begin{cases} 2^r, & r \text{ even} \\ -2^r, & r \text{ odd} \end{cases}$$

CHAPTER 4

Graph Theory

DEFINITION

A linear graph G is a pair (V, E) consisting a set of objects $V = \{v_1, v_2, ...\}$ called vertices and another set $E = \{e_1, e_2, ...\}$, whose elements are called edges, such that each edge is identified with a pair (v_i, v_j) of vertices. The vertices v_i, v_j associated with edge e_k are called the end-vertices of e_k. We generally represent a graph by means of a diagram, in which the vertices are represented as points and each edge as a line segment joining its end vertices. Often this diagram itself is referred to as the graph.

An edge can be associated with a vertex pair (v_i, v_j). An edge having the same vertex as both its end-vertices is called a self-loop. There can be more than one edge associated with a given pair of vertices. Such edges are referred to as parallel edges. A graph that has neither self-loops nor parallel edges is called a simple graph. The term general graph is used to mean that parallel edges and self-loops are allowed.

When a vertex v_i is an end vertex of some edge e_j; v_i and e_j are said to be incident with each other. Two non-parallel edges are said to be adjacent if they are incident on a common vertex, or if they are the end vertices of the same edge.

The number of edges incident on a vertex v_i, with self loop counted twice, is called the degree, $d(v_i)$, of the vertex v_i. The degree of a vertex is sometimes also referred to as its valency.

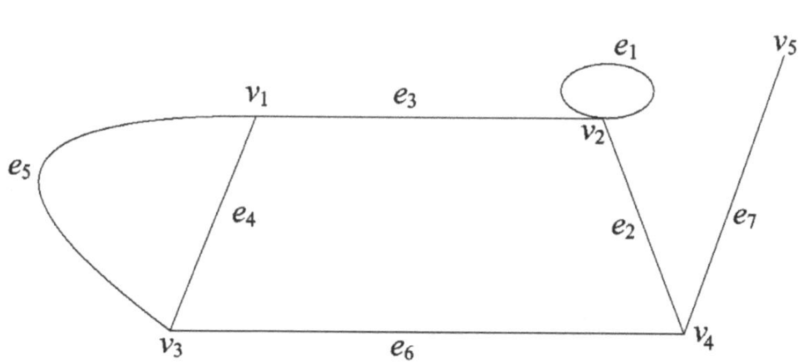

Figure 4.1

The Figure 4.1 represents a graph with five vertices and seven edges. In this case:

e_1 is a self-loop
e_4 and e_5 are parallel edges
e_2, e_6 and e_7 are incident with v_4
e_2 and e_7 are adjacent
v_4 and v_5 are adjacent
v_1 and v_4 are not adjacent

$$d(v_1) = d(v_3) = d(v_4) = 3$$
$$d(v_2) = 4$$
$$d(v_5) = 1.$$

4.1 FINITE AND INFINITE GRAPHS

A graph whose vertex set *V* as well as the edge set *E* are finite is called a finite graph; otherwise the graph is called an infinite graph.

Null Graph

A graph $G = (V, E)$ is said to be null graph if $E = \Phi$. Some authors define null graph to be a graph $G(V, E)$ with $V = \Phi$ and $E = \Phi$ and they call a graph with $E = \Phi$ and $V \neq \Phi$ a vertex graph. In fact, if $V = \Phi$ and $E = \Phi$, then there is no graph and we agree that a graph must have at least one vertex.

Complete Graph

A complete graph is a simple graph in which there exists an edge between every pair of vertices.

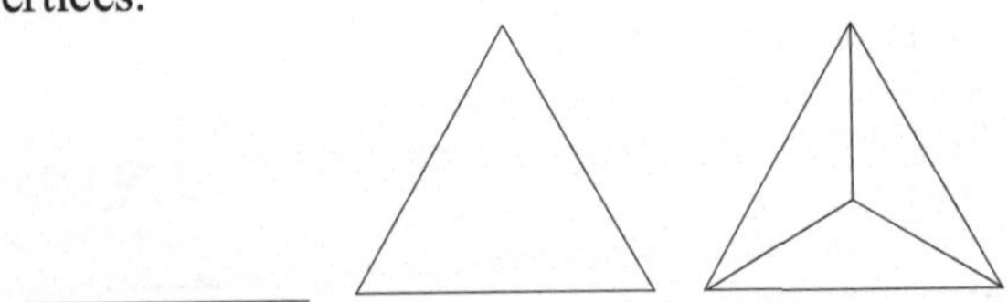

Paths and Circuits

Isomorphism

Two graphs G and G' are said to be isomorphic to each other if there is a one-one correspondence between their vertices and between their edges such that the incidence relation is preserved. In other words, if the edge e is incident on vertices v_1 and v_2 of G, then the corresponding edge e' in G' must be incident on vertices v_1' and v_2' that correspond to v_1 and v_2 respectively.

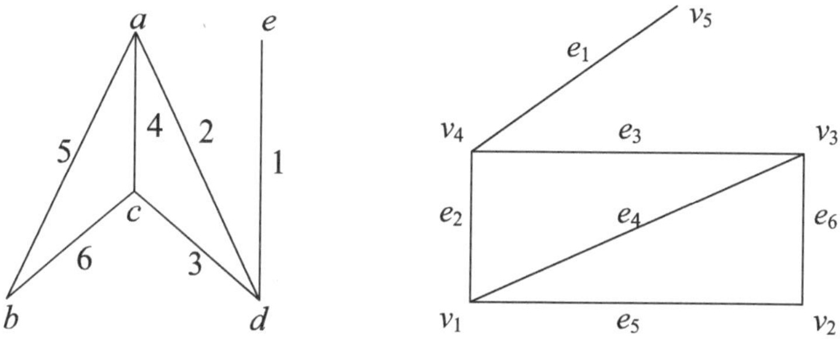

The above two graphs are isomorphic to each other. In this case the vertices a, b, c, d, e correspond to v_1, v_2, v_3, v_4, v_5 respectively and the edges $1, 2, 3, 4, 5, 6$ correspond to $e_1, e_2, e_3, e_4, e_5, e_6$ respectively.

Subgraph

A graph g is said to be a subgraph of a graph G if all the vertices and all the edges of g are in G, and each edge of g has the same end vertices in g as in G. For example, of the following two graphs A is a subgraph of B. Subgraph is same as the concept of a subset in set theory. The reader may note the following facts about subgraphs.

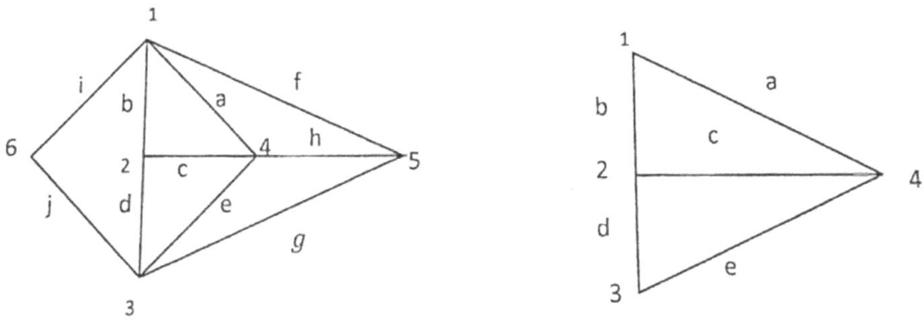

1. Every graph is a subgraph of itself.
2. A subgraph of a subgraph of G is also a subgraph of G.
3. A single vertex in a graph G is a subgraph of G.
4. A single edge in G, together with its end vertices, is a subgraph of G.

Walks, Paths and Circuits

A walk is defined as a finite alternating sequence of vertices and edges, beginning and ending with vertices, such that each edge is incident with the vertices preceding and following it. No edge appears more than once in a walk; a vertex, however, may appear more than once.

Vertices with which a walk begins and ends are called its terminal vertices. It is possible for a walk to begin and end at the same vertex. Such a walk is called a closed walk. A walk that is not closed is called an open walk. Note that for an open walk the terminal vertices are distinct.

An open walk in which no vertex appears more than once is called a path. A path does not intersect itself. The number of edges in a path is called the length of a walk. It follows that an edge which is not self loop is a path of length one. A self-loop can be included in a walk but not in a path. The terminal vertices of a path are of degree one and the rest of the vertices are of degree 2.

A closed walk in which no vertex, except the initial and final vertex, appears more than once is called a circuit. In other words, a circuit is a closed non-intersecting walk.

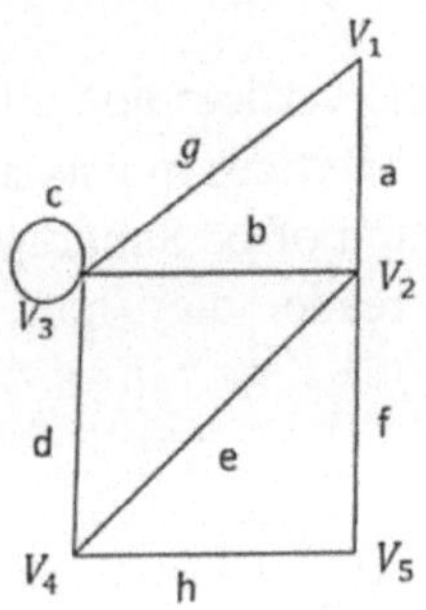

An open walk

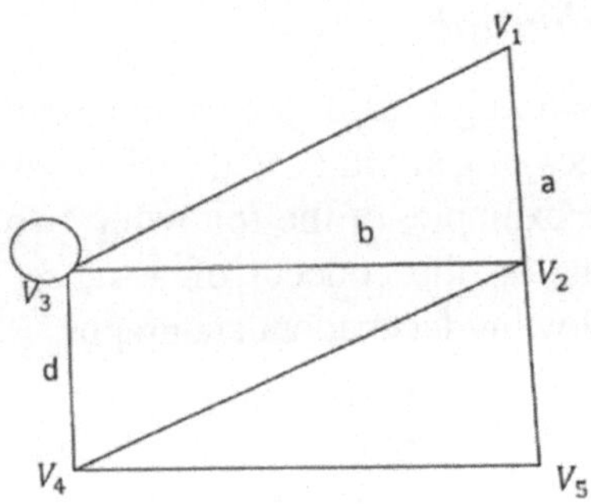

Path of length three

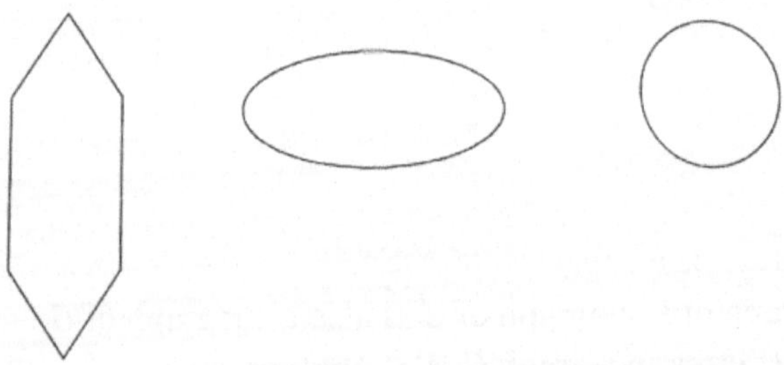

A graph is connected if we can reach any vertex from any other vertex by travelling along the edges. To be more precise, a graph G is said to be connected if there is atleast one path between every pair of vertices in G. Otherwise, G is disconnected.

Each connected subgraph is called a component.

A simple graph in which there exists an edge between every pair of vertices is called a complete graph.

A tree is a connected graph without any circuits. Note that it has to be simple graph having neither a self-loop nor parallel edges.

If some closed walk in a graph contains all edges of the graph, then the graph is called an Euler graph.

4.2 OPERATIONS ON GRAPHS

Let $G_1 = (V_1, E_1)$ and $G_2 = (V_2, E_2)$ be any two graphs. Then the union of G_1 and G_2(written as $G_3 = G_1 \cup G_2$) is the graph $G_3 = (V_3, E_3)$ whose vertex set $V_3 = V_1 \cup V_2$ and the edge set $E_3 = E_1 \cup E_2$. The intersection $G_1 \cap G_2$ of G_1 and G_2 is the graph $G_4 = (V_4, E_4)$ where $V_4 = V_1 \cap V_2$ and $E_4 = E_1 \cap E_2$. The ring sum of G_1 and G_2 (written as $G_1 \oplus G_2$) is a graph consisting of the vertex set $V_1 \cup V_2$ and of edges that are either in G_1or in G_2, but not in both.

The following figure shows the example of two graphs G_1 and G_2, their union, intersection and the ring sum.

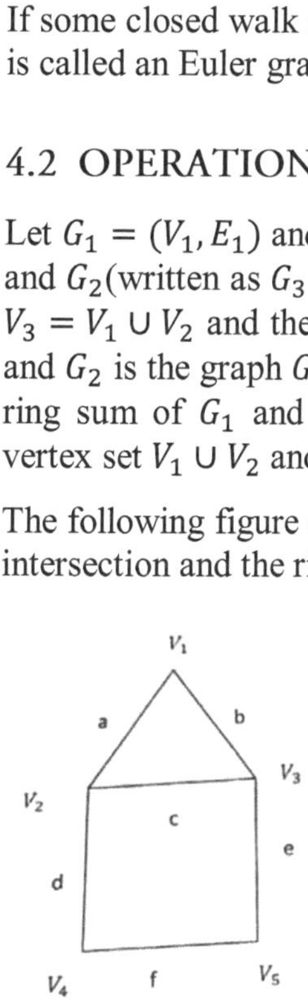

G_1

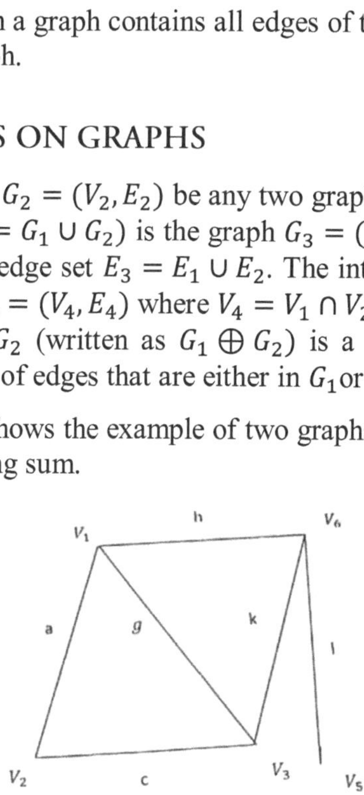

G_2

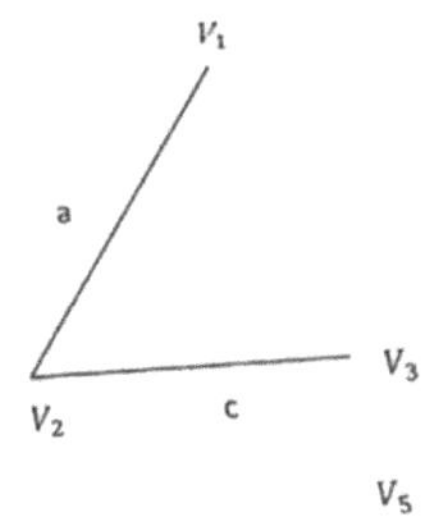

$G_1 \cap G_2$

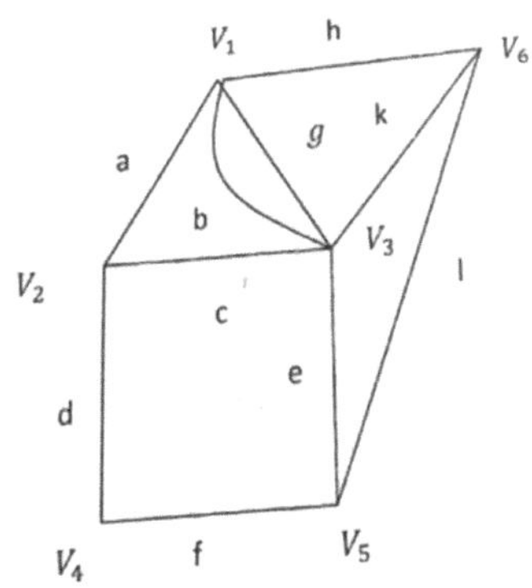

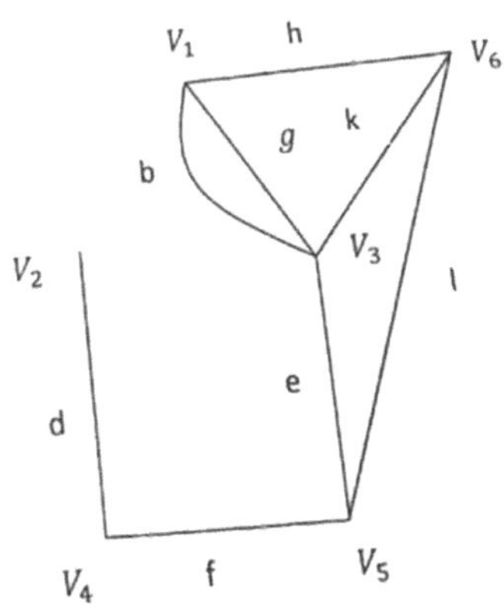

Union, intersection and the ring sum are commutative, i.e. for any two graphs G_1 and G_2 we have

$$G_1 \cup G_2 = G_2 \cup G_1, G_1 \cap G_2 = G_2 \cap G_1, G_1 \oplus G_2 = G_2 \oplus G_1$$

Also for any graph G we have

$$G \cup G = G = G \cap G$$

and $G \oplus G$ = a null graph.

Decomposition

A graph G is said to have been decomposed into two subgraphs g_1 and g_2 if

$g_1 \cup g_2 = G$ and $g_1 \cap g_2 = a$ null graph.

In other words, every edge of G occurs either in g_1 or in g_2, but not in both; but some of the vertices may occur in both g_1 and g_2.

Fusion

A pair of vertices a and b in a graph are said to be fused (identified or merged) if the two vertices are replaced by a single new vertex such that every edge that was incident on either a or b or on both is incident on the new vertex. Thus, fusion of two vertices does not alter the number of edges, but reduces the number of vertices by one.

The number of edges incident on a vertex v_i of a graph, with self-loops counted twice, is called the degree of the vertex v_i and is denoted by $d(v_i)$.

In Figure 4.1 $d(v_1) = d(v_3) = d(v_4) = 3, d(v_2) = 4, d(v_5) = 1$.

Let G be a graph with e edges and n vertices $v_1, v_2, \dots, v_n$.

Since each edge contributes two degrees, the sum of the degrees of all vertices in G is twice the number of edges in G. In mathematical notations

$$\sum_{i=1}^{n} d(v_i) = 2e$$

In Figure 4.1

$$\begin{aligned} d(v_1) + d(v_2) + d(v_3) + d(v_4) + d(v_5) \\ &= 3 + 4 + 3 + 3 + 1 \\ &= 14 \\ &= \text{twice the number of edges.} \end{aligned}$$

We have the following Theorems

4.2.1 *Theorem:* In any graph, the number of vertices of odd degree is even.

Proof: Let $v_1, v_2, \dots, v_n$ be the vertices of a graph G. Then

$$\sum_{i=1}^{n} d(v_i) = \sum_{\text{even}} d(v_j) + \sum_{\text{odd}} d(v_i).$$

The L.H.S. of the above equation is even, the first term on R.H.S. is even. Hence, the second term must also be even, i.e.,

$$\sum_{\text{odd}} d(v_i) = \text{an even integer.}$$

Since each $d(v_k)$ is odd, the total number of terms in the sum must be even and this completes the proof.

4.2.2 *Theorem:* A graph G is disconnected if and only if its vertex set V can be partitioned into two nonempty, disjoint subsets V_1 and V_2 such that there exists no edge in G whose one end vertex is in V_1 and the other in V_2.

Proof: Let G be a disconnected graph and let a be a vertex in G. Let V_1 be the set of all vertices that are joined by paths to a. Since G is disconnected, V_1 does not include all vertices in G. The remaining vertices will form a nonempty set V_2. No vertex in V_1 is joined to any vertex in V_2 by an edge and hence, this is a partition.

Conversely, suppose that such a partition exists. Let a and b be arbitrary vertices in G such that $a \in V_1$ and $b \in V_2$. We claim that no path exists between a and b, for otherwise there would be atleast one edge whose one end vertex would be in V_1 and the other in V_2. Hence G is not connected.

4.2.3 *Theorem:* If a graph has exactly two vertices of odd degrees there must be a path joining these two vertices.

Proof: Let G be a graph with all even vertices except V_1 and V_2 which are odd. But no graph can have an odd number of odd vertices. Therefore, V_1 and V_2 must belong to the same component and thus, must have a path between them.

The proof of the next theorem requires the following Lemma (inequality).

Lemma

$$\sum_{i=1}^{k} n_i^2 \leq n^2 - (k-1)(2n-k).$$

Proof: We know that,

$$\sum_{i=1}^{k}(n_i - 1) = n - k.$$

Squaring both sides,

$$\left(\sum_{i=1}^{k}(n_i - 1)\right)^2 = n^2 + k^2 - 2nk$$

or,

$\sum_{i=1}^{k}(n_i^2 - 2n_i) + k +$ non-negative cross terms $= n^2 + k^2 - 2nk$ because $(n_i - 1) \geq 0$ for all i.

Therefore,

$$\sum_{i=1}^{k} n_i^2 \leq n^2 + k^2 - 2nk - k + 2n = n^2 - (k-1)(2n-k).$$

4.2.4 *Theorem:* If a graph has exactly two vertices of odd degree, there must be a path joining these two vertices.

Proof: Let the number of vertices in each of the k-components of a graph G be $n_1, n_2, \dots, n_k$. Now we have,

$$n_1 + n_2 + \cdots + n_k = n, n_i \geq 1.$$

The maximum number of edges in the ith component of G (which is a simple connected graph) is $n_i(n_i - 1)$. Hence, the maximum number of edges in G,

$$= \frac{1}{2}\sum_{i=1}^{k} n_i(n_i - 1) = \frac{1}{2}\left(\sum_{i=1}^{k} n_i^2\right) - \frac{n}{2}$$

$$\leq \frac{1}{2}[n^2 - (k-1)(2n-k)] - \frac{n}{2}$$

$$= \frac{1}{2}(n-k)(n-k+1)$$

We now discuss an important and interesting example where graph theory is applied. This is known as Konigsberg Bridge problem.

The Konigsberg Bridge problem is one of the best known problems which remained unsolved for a long time. Finally it was solved by Euler (1707–1783) in the year 1736 by using graph. In fact graph theory originated from this problem and hence, Euler can be regarded as the founder of graph theory. We shall now state the problem and then discuss its solution. We begin with the diagram. Two islands C and D formed by the river Pregel in

Konigsberg (then the capital of East Prussia, now renamed Kalingard) were connected to each other and to banks A and B by seven bridges as shown in the figure below. The problem was to start at any of the four land areas A, B, C, or D, walk over each of the seven bridges exactly once and return to the starting point.

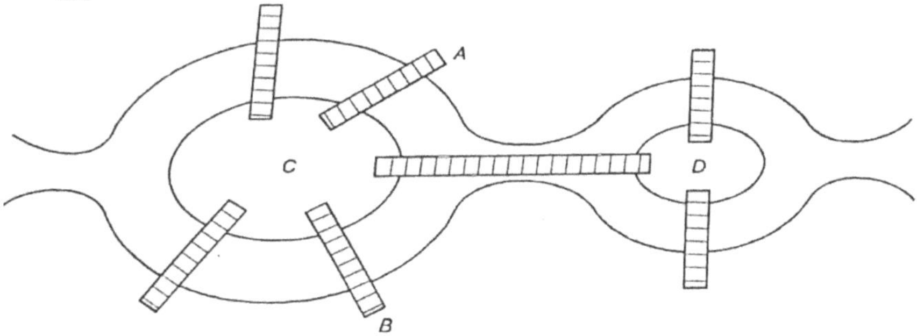

Euler represented this problem by means of a graph as shown below, where the vertices represent the land areas and the edges represent the bridges. Observe that not all its vertices are of even degrees. Hence, it is not an Euler graph. Therefore it is not possible to walk over each of the seven bridges exactly once and return to the starting point. In other words, it means that the solution to the Konigsberg Bridge problem does not exist.

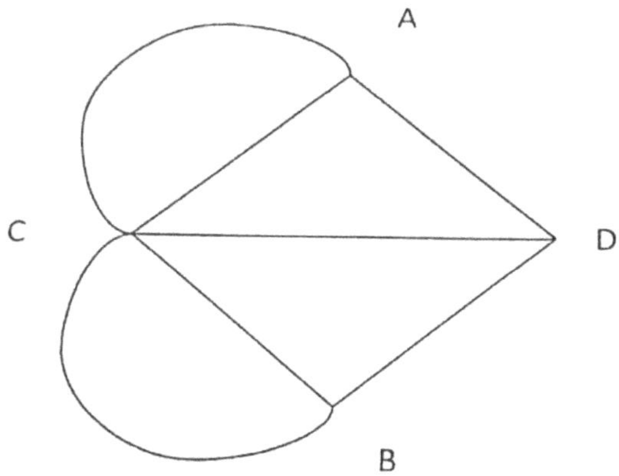

Recall that a tree is a connected graph without any circuits. We now state a theorem which characterises a tree.

4.2.5 *Theorem:* A graph G is a tree if and only if it has only one path between every pair of vertices.

***Proof*:** Let G be a tree. Since G is connected, there must exist at least one path between every pair of vertices in G. Suppose there are two distinct paths between two vertices a and b of G. The union of these two paths will contain a circuit and hence G can not be a tree.

Conversely, let there be only one path between every pair of vertices in G. Hence, G is connected. Observe that G has no circuit, for otherwise, there would be at least one pair of vertices a, b such that there would be two distinct paths between a and b. Therefore, G is a tree and this completes the proof.

Besides solving the Konigsberg bridge problem, Euler posed and solved a more general problem. The problem he posed was to characterise those graphs for which it is possible to find a closed walk running through every edge exactly once. Such walks are now-a-days called Euler lines and a graph that consists of an Euler line is called an Euler graph. More precisely, an Euler line is a closed walk in a graph which contains all the edges of the graph and the graph is called an Euler graph.

It should be noted that an Euler graph is connected. We have the following theorem which is required to show that Konigsberg Bridge problem has no solution. We shall only state it here.

4.2.6 *Theorem:* A connected graph G is an Euler graph if and only if all vertices of G are of even degree.

We shall now discuss some basic properties of tree.

4.2.7 *Theorem:* A tree with n vertices has $(n-1)$ edges.

Proof: We shall prove this theorem by induction on n, the number of vertices.

It is easy to see that it is true for $n = 1$. Assume that the theorem holds good for all trees with fewer than n vertices.

Now consider a tree T with n vertices. Let e_k be an edge in T with end vertices v_i and v_j. From Theorem 4.2.5 it follows that there is no other path between v_i and v_j except e_k. Hence, If we delete e_k from T, then T will be disconnected. Also $T - e_k$ consists of exactly two components; since there is no circuit on T to start with, each of these components is a tree. Both these trees t_1 and t_2 have fewer than n vertices and therefore, by the induction hypothesis, each contains one edge less than the number of vertices in it. Thus, $T - e_k$ consists of $(n-2)$ edges and hence, T has $(n-1)$ edges and this completes the proof.

Before we proceed to the next result we shall introduce a definition.

4.2.1 *Definition:* A connected graph is said to be minimally connected if the removal of any one edge from it makes the graph disconnected.

4.2.8 *Theorem:* A graph is a tree if and only if it is minimally connected.

Proof: Let T be a minimally connected graph. T cannot have a circuit, otherwise removal of one of the edges in the circuit will leave T connected. Hence, T is a tree.

Conversely, let T be a connected graph and not minimally connected. Then there exists an edge e_i in T such that $T - e_i$ is connected. Therefore, e_i is in some circuit and this implies that T is not a tree.

4.2.9 *Theorem:* A graph G with n vertices, $n - 1$ edges and no circuit is connected.

Proof: Suppose, on the contrary, there exists a graph G with n vertices, $n - 1$ edges and no circuits and it is disconnected. Then, G consists of two components with no circuits. To be specific, let G consist of two components g_1 and g_2 each with no circuits. Now add an edge e between a vertex $v_1$in g_1 and v_2 in g_2. Since there was no path between v_1 and v_2 in G, addition of e does not give rise to a circuit. Hence, $G \cup e$ is a connected graph with no circuits (i.e., a tree) with n vertices has $n - 1$ edges.

Thus, observe that the following conditions are equivalent:

1. T is a tree,
2. T is connected with no circuit,
3. T is connected and has $n - 1$ edges,
4. T has no circuit and has $n - 1$ edges,
5. There is exactly one path between every pair of vertices in T,
6. T is minimally connected.

We shall now discuss what is called a planar graph.

4.2.2 *Definition:* A graph G is called a planar graph if there exists some geometrical representation of G which can be drawn on a plane such that no edge intersect except at a vertex.

4.2.10 *Theorem:* For any connected planar graph,

$$v - e + r = 2(*)$$

where v, e and r are the number of vertices, edges and regions of the graph, respectively.

Proof: We shall prove this theorem by induction on the number of edges. Observe that the theorem is true for the two graphs with a single edge.

Assume that the result is true for all graphs with $n-1$ edges. Let G be a connected graph with n edges. If G has a vertex of degree 1, the removal of this vertex together with the edge incident with it will yield a connected graph G' in which the property $(*)$ holds. Therefore $(*)$ also holds in G, because putting the removed edge and vertex back into G will increase the number of vertices by 1, number of edges by 1 but not change the count of the regions. If G has no vertex of degree 1, the removal of any edge in the boundary of a finite region will yield a connected graph G' in which $(*)$ is satisfied. Hence $(*)$ will be satisfied in G, because putting the removed adge back into G will increase the number of edges by 1, number of regions by 1 but not change the count of vertices.

This completes the proof.

4.3 GRAPH-THEORETIC ALGORITHMS

In this section we shall discuss some graph-theoretic algorithms. For this purpose we need two most important matrix representations of graphs and discuss these concepts first.

Incidence Matrix

Let G be graph having n veritces, e edges and without self-loops. We define an $n \times e$ matrix $A = (a_{ij})$ as follows:

$$a_{ij} = \{1 \text{ if } j\text{th edge } e_j \text{ is incident on the } i\text{th vertex } v_i, 0 \text{ otherwise}\}.$$

The n rows of this matrix correspond to the n vertices and the e columns to the e edges of the graph. This matrix A is called the vertex-edge incident matrix or simply the incident matrix of the graph G and is written as $A(G)$. This is a matrix of 0's and 1's only and such matrices (matrices with zeros and 1's only) are called binary matrices.

Some of the basic properties of an incident matrix A can be readily observed as follows:

1. Since every edge is incident on exactly two vertices, each column of A has exactly two 1's.
2. The number of 1's in each row equals the degree of the corresponding vertex.
3. A row with all 0's represents an isolated vertex.
4. Parallel edges in a graph produce identical columns in its incidence matrix.

Adjacency Matrix

Let G be a graph with n vertices and no parallel edges. The adjacency matrix of the graph G is an $n \times n$ symmetric binary matrix $X = (x_{ij})$ defined, over the ring of integers, as follows:

$$x_{ij} = \{1 \text{ if there is an edge between } i\text{th and } j\text{th vertices}, \\ 0 \text{ if there is no edge between them}\}$$

Some of the basic properties of the adjacency matrix are as follows:

1. The entries along the principal diagonal of X are all 0's if the graph has no self-loop. A self-loop at the ith vertex corresponds to $x_{ii} = 1$.
2. The definition of adjacency matrix makes no provision for parallel edges. This is why the adjacency matrix X is defined for graphs without parallel edges.
3. If the graph has no self-loops (and no parallel edges), the degree of a vertex is equal to the number of 1's in the corresponding row or column of X.

Algorithms

An algorithm is a set of instructions which follows step by step to provide the solution of the given problem. Each step of the algorithm has to be precisely defined and the algorithm should terminate after a finite number of steps. Usually an algorithm is first written in an ordinary language, (for example in English language for our purpose) and then it is into a flow chart and finally, it is written in a computer language so that the machine can understand and execute it. Efficiency of an algorithm depends on the memory and computation-time requirement as a function of the size of the input. When the input is a graph its size is the number of vertices and number of edges. For the efficiency of graph, theoretic algorithms, memory requirement is not a problem but the computation time could be when graph is the input, it is represented in some matrix forms. We shall now discuss this in the following section.

Input

1. In most cases the graph is fed to a computer by its adjacency matrix. After assigning a distinct number to each of the n vertices of the graph G, then $n \times n$ binary matrix $x(G)$ is used to represent G during input, storage and output. Since each of the n^2 entries in the adjacency matrix is either 0 or 1, we require n^2 bits in the computer memory. Bits can be

packed into words. Let w be the word length i.e., the number of bits in a computer word. Let n be the number of vertices in the graph. Then, each row of the adjacency matrix can be written as a sequence of n bits in $[n\backslash w]$ machine words, where $[x]$ denotes the smallest integer not less than x. Therefore, the number of words required to store the adjacency matrix is $n[n\backslash w]$.

2. *Incidence matrix*: Sometimes an incidence matrix also used to store and manipulate a graph in a computer. An incidence matrix requires n bits of storage, which may be more than n^2 bits needed for an adjacency matrix, because the number of edges e is greater than the number of vertices n.

A Basic Algorithm

We shall now discuss a basic graph-theoretic algorithm called connectedness and components.

Connectedness and Components

We start with some vertex in the graph G and fuse all vertices that are adjacent to it. We then take the fused vertex and again fuse with it all those vertices that are adjacent to it. This process of fusion is repeated until no more vertices can be fused. This indicates that a connected component has been fused to a single vertex. If all the vertices are exhausted, then the graph is connected. Otherwise, start with a new vertex belonging to a different component and continue to fuse as before.

In the adjacency matrix the fusion of the jth vertex to the ith vertex will be accomplished by logically adding the jth row to the ith row and also the jth column to the ith column. (Logical addition means $1 + 0 = 0 + 1 = 1 + 1 = 1$ and $0 + 0 = 0$) then the jth row and jth column will have to be discarded from the matrix. A self-loop resulting from a fusion appears as 1 in the main diagonal but parallel edges are replaced by a single edge because of the logical addition operation and these have no effect on the connectedness of the graph.

If n is the number of vertices, the maximum number of fusions that may be necessary to perform in the algorithm is $n - 1$ and in each fusion at most n logical additions are performed. Therefore the upper bound on the execution time is proportional to $n(n - 1)$. A proper choice of the initial vertex in each component will improve the efficiency. A flow chart is given below.

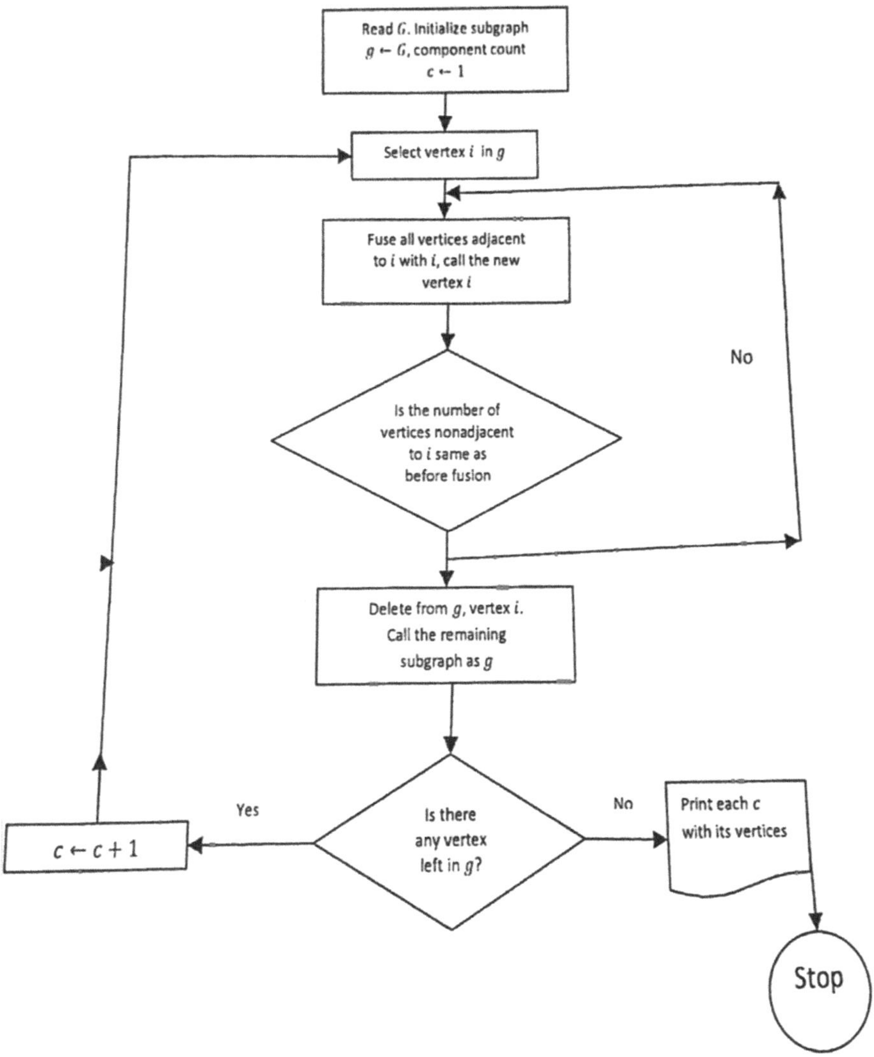

EXERCISES

1. Show that the maximum degree of any vertex in a simple graph with n vertices is $n-1$.
2. Show that the maximum number of edges in a simple graph with n vertices is $\frac{n(n-1)}{2}$.
3. Prove that a simple graph with n vertices must be connected if it has more than $\frac{(n-1)(n-2)}{2}$ edges.

CHAPTER 5

Group Theory

5.1 DEFINITIONS AND EXAMPLES

Definitions

Let G be a non empty set. Let there be a mapping from $G \times G$ into G called a *binary composition*, defined by,

$$(a, b) \to ab$$

Then G along with this binary composition is called a group if it *satisfies* the following properties:

1. For $a, b \in G$ we have $ab \in G$ (closure property),
2. For $a, b, c \in G$ we must have

   $a(bc) = (ab)c$ (associative law)
3. There exists an element $e \in G$, called the identity, such that for every $a \in G, ae = ea = a$ (the existence of an identity),
4. For every $a \in G$, there exists an element $a^{-1} \in G$ called the inverse of a in G such that,

   $aa^{-1} = a^{-1}a = e$ (the existence of inverse)

   The property (i) is a superfluous statement, it only tells again that the mapping $(a, b) \to ab$ is from $G \times G$ into G, i.e. it is a binary composition. G with property (i) only is called a *groupoid*, with (i) and (ii) is called a semigroup and with (i), (ii) and (iii) only is called a *monoid.*

   A group G is called *abelian* or *commutative* if it has the following property.
5. For $a, b, \in G, ab = ba$ (commutative law). Otherwise a group G is called *non-abelian.*

A group G is called finite if it is a finite set, otherwise it is called an *infinite group.* The number of elements of a group G is called the *order* of G and is denoted by $o(G)$. This number is most interesting, useful and meaningful when G is a finite group.

Remark

In the above definitions we have used multiplicative notation for the binary composition. If one uses additive notation the definition of a group reads as follows.

A non empty set G along with a mapping $+: G \times G \to G$ defined by

$$(a, b) \to a + b$$

Is called a group if:

1. $a, b \in G$ implies $a + b \in G$
2. $a, b, c \in G$ implies $a + (b + c) = (a + b) + c$
3. For every $a \in G$, there exists an element $0 \in G$ such that,

   $a + 0 = 0 + a = a$

4. For every $a \in G$, there exists an element $(-a) \in G$ such that:

   $a + (-a) = (-a) + a = 0$

However while proving the general results we shall use the multiplicative notation for the binary composition. Before we go further we shall first give some examples of groups leaving the verification for the reader as exercise.

Examples of Groups

1. With respect to usual addition of numbers the following sets form groups.
   (a) The singleton set consisting only of the number 0.
   (b) The set **Z** of all integers. The set of all nonnegative integers does not form a group.
   (c) The set of all even integers including 0. Observe that the set of all odd integers is not a group.
   (d) The set **Q** of all rational numbers.
   (e) The set **R** of all real numbers.
   (f) The set **C** of all complex numbers.
   (g) The set of all complex numbers, whose real and imaginary parts are integers.
2. The following sets of numbers form groups with respect to usual multiplication of numbers.
   (a) The singleton set consisting only of the number 1.
   (b) The set $\{1, -1\}$.
   (c) $\{1, -1, i, -i\}$.
   (d) The set **Z** of all integers is not a group.

(e) The set of all positive rational numbers (excluding 0). The set of all nonzero rationals. Observe that the set of all negative rationals is not a group.
(f) The set of all positive real numbers. Note that the negative real numbers do not form a group.
(g) The set of all nonzero real numbers. The set **R** of all real numbers does not form a group.
(h) The set of all nonzero complex numbers. Observe that the set **C** of all complex numbers does not form a group.
(i) The unit circle $\{z \in C : |z| = 1\}$. This is called the circle group.
(j) The set $\{z \in C : z^n = 1\}$ where n is a fixed but arbitrary positive integer.

3. The set **R**" of all n-tuples of real numbers, i.e.,
$\mathbf{R}" = \{x = (x_1, ..., x_n) : x_i \in \mathbf{R}, i = 1, \ldots, n\}$,
Is a group with respect to coordinate wise addition defined by,
$x + y = (x_1 + y_1, ..., x_n + y_n)$
Where $x = (x_1, ..., x_n)$ and $y = (y_1, ..., y_n)$. Similarly the set **C**" of all n-tuples of complex numbers forms a group with respect to coordinate wise addition.
4. The family of all subsets of a set U is a group with respect to the formation of symmetric difference of two sets A and B defined by,
$A \Delta B = (A - B) \cup (B - A)$
It should be noted that if $U \neq \varphi$, then the above family of sets does not constitute a group with respect to the formation of either unions of intersections.
5. Let m be a positive integer and let $\mathbf{Z}_m$, be the set of all nonnegative integers less than m, i.e. $\mathbf{Z}_m = \{0, 1, \ldots, m - 1\}$. If a and b are in $\mathbf{Z}_m$ define their sum $a + b$ to be the remainder obtained when their usual sum is divided by m. Then $\mathbf{Z}_m$ is a group with respect to the operation defined above.
6. The set of all 1–1 mappings of a nonempty set X onto itself is a group with respect to the multiplication (composition) of mappings defined as follows: if f and g are two such mappings and x is an arbitrary element of X then $(fg)(x) = f(g(x))$.
7. As a special case of the above example let $X = \{1, 2, ..., n\}$. This case is of particular importance. A 1–1 mapping of this set X onto itself is called a permutation and can be written in the convenient form,
$\begin{pmatrix} 1 & 2 \ldots & n \\ i_1 & i_2 \ldots & i_n \end{pmatrix}$

where below each of the integers $1, 2, \dots, n$ is placed its image under the mapping.

For example, if $n = 4$ two such mappings p and q can be conveniently written in the form,

$$p = \begin{pmatrix} 1 & 2 & 3 & 4 \\ 3 & 1 & 4 & 2 \end{pmatrix},\ q = \begin{pmatrix} 1 & 2 & 3 & 4 \\ 4 & 1 & 3 & 2 \end{pmatrix}$$

We define the product pq of p and q by,

$$\begin{pmatrix} 1 & 2 & 3 & 4 \\ 2 & 3 & 4 & 1 \end{pmatrix}$$

The set of all permutations of n elements is denoted by S_n and this forms a group under the product of permutations. This group is called the symmetric group of degree n.

8. Consider the set of all $m \times n$ (with m and n fixed positive integers) matrices with real or complex entries. If $A = (a_{jk})$ and $B = (b_{jk})$ are two such matrices then we define the sum $A + B$ to be the matrix $C = (c_{jk})$ with elements,

   $c_{jk} = a_{jk} + b_{jk}, j = 1, \dots, m, k = 1, \dots, n$

   The above set forms a group with respect to the operation of matrix addition defined as above. In particular the set of $n \times n$ square matrices forms a group under matrix addition, but this set does not form a group under matrix multiplication. Any n-rowed square matrix A need not have an inverse A^{-1} such that,

   $AA^{-1} = A^{-1}A = I$

   where I is the n-rowed identity (unit) matrix. This holds if and only if A is nonsingular, that is, if and only if $detA \neq 0$ where $detA$ denotes the determinant A.

9. The set of all n^{th} roots of 1 is a group with respect to multiplication. If A denotes the set of all n^{th} roots of 1 then,

   $A = \{e^{2\pi ir/n} : r = 0, 1, 2, \dots, (n-1)\}$

   If $a = e^{2\pi_i r_1/n}, 0 \leq r_1 \leq n-1$ and $b = e^{2\pi_i r_2/n},\ 0 \leq r_2 \leq n-1$, then $ab = e^{2\pi_i(r_1+r_2)/n}$. If $r_1 + r_2 \leq n-1$, then $ab \in A$. If $r_1 + r_2 \geq n$ then assume that $r_1 + r_2 = tn + k, 0 \leq k \leq n-1$ and then $ab = e^{2\pi i(tn+k)/n} = e^{2\pi it} . e^{2\pi ik/n} = e^{2\pi ik/n} \in A$ since $0 \leq k \leq n-1$.

   The verification of other axioms of group are left to the reader. It should be noted that this group is abelian, for

   $e^{2\pi ir_1/n} . e^{2\pi ir_2/n} = e^{2\pi i(r_1+r_2)/n} = e^{2\pi i(r_1+r_2)/n} = e^{2\pi ir_2/n} . e^{2\pi ir_1/n}$

Worked Out Examples

1. Show that if every element of the group G is its own inverse, then G is abelian.

   Let $a, b \in G$. Then $a^{-1}, b^{-1} \in G, ab \in G$ and $(ab)^{-1} \in G$. But $a = a^{-1}, b = b^{-1}$. Now $ab = (ab)^{-1} = b^{-1}a^{-1} = ba$.

   Hence G is abelian.

   For any positive integer n we write $a^n = aaa \dots a$ (n times).

2. If $a^2 = e$ for every element a in a group G, then show that G is abelian.

   We have $a^2 = aa = e.$ But $aa^{-1} = e.$

   Therefore $aa = e = aa^{-1}$ and so $a = a^{-1}$ for every $a \in G$. Now it follows from the previous example that G is abelian.

3. If a group G has four elements, then show that it must be abelian.

   Let $G = \{e, a, b, c\}$ where e is the unit element in G.

   If each element is its own inverse, then G is abelian. If not, assume that $a^{-1} = b$ and $c^{-1} = c$.

   That is, $ab = ba = e$ and $cc = e$. Now

   $ac \neq e$ since $a^{-1} = c$

   $ac \neq a$ since $c \neq e$

   $ac \neq c$ since $a \neq e$

   Therefore $ac = b$. Similarly it can be shown that $ca = b.$ Thus $ac = ca$.

   $bc \neq e$ since $b^{-1} \neq c$

   $bc \neq b$ since $c \neq c$

   $bc \neq c$ since $b \neq e$

   Therefore $bc = a$ and similarly $cb = a$ so that $bc = cb$. Thus G is abelian. The binary composition in G which we call multiplication, is given by the following table,

	e	a	b	c
e	e	a	b	c
a	a	c	e	b
b	b	e	c	a
c	c	b	a	e

4. If G is a group such that $(ab)^2 = a^2b^2$ for all $a, b \in G$ show that G must be abelian.

for all $a, b \in G$ we have,
$(ab)^2 = a^2b^2$
i.e. $(ab)(ab) = (aa)(bb)$
i.e. $a(ba)b = a(ab)b$
i.e. $a(ba) = a(ab)$ (by right cancellation)
i.e. $ba = ab$ (by left cancellation)
Therefore G is abelian.

5.1.1 *Definition*

If G is a group and $a \in G$ then the order of a is defined to be the least positive integer n such that $a^n = e$. If there exists no such positive integer then a is said to be of infinite order. We write $o(a)$ to denote the order of a.

5.1.1 *Example*

1. In any group $G, o(e) = 1$, where e denotes the identity element in G.
2. In the multiplicative group $\{1, -1, i, -i\}, o(-1) = 2, o(i) = 4$.
3. In the group
   $\{\ldots, a^{-3}, a^{-2}, a^{-1}, a^0 = \text{e}, a, \text{a}^2, \text{a}^3, \ldots\}$
   the order of any element is either 0 or ∞.
4. If w denotes a cube root of 1 then in the group,
   $\{1, \omega, \omega^2\}, o(1) = 1, o(\omega) = o(\omega^2) = 3$

5.1.1 *Theorem*

1. Let G be a group and $a \in G$. Then for any positive integer $k, o(a^k) \leq o(a)$.
2. $o(a^{-1}) = o(a)$.

   Proof: (i) Let $o(a) = n$ and $o(a^k) = m$. Therefore $a^n = e$ and $(a^k)^m = e$. So $(a^k)^n = (a^n)^k = e^k = e$. But m is the smallest positive integer for which $(a^k)^m = e$. So $m \leq n$.
   Similarly,
   $a^m = ((a^{-1})^{-1})^m = (a^{-1})^{-m} = e^{-1} = e$ and so $n \leq m$.
   Therefore, $m = n$ and this completes the proof.

5.1.2 *Theorem*

The order of every element of a finite group G is finite.

Proof: Let $o(G) = n$ and let $a \in G$ be arbitrary. The elements $e, a, a^2, \ldots,$ a^n are $(n+1)$ innumber and $o(G) = n$ and hence they are not all distinct. In other words, there exist r, s with $1 \leq r \leq s \leq n$ such that $a^r = a^s$. Hence $e = a^r a^{-r} = a^s a^{-r} = a^{s-r} = a^k$ where $k = s - r, 1 \leq k \leq n.$

Therefore, $o(a)$ is finite and since $a$is arbitrary, this completes the proof.

5.1.2 *Definition*

Let G be a group. If for some $a \in G$ each element of G is precisely of the form a^n for some $n \in Z$ then G is said to be a cyclic group and a is said to be generator of G. Note that if G is a cyclic group and $a$is its generator, then

$$G = \{a^n : n \in Z\} = (a)$$

5.1.2 *Example*

(a) Under addition,

$$Z = \{\ldots, -2, -1, 0, 1, 2, \ldots\}$$

is a cyclic group with 1 as the generator. Also under addition,

$$\{\ldots, -4, -2, 0, 2, 4, \ldots\}$$

is a cyclic group with 2 as generator.

(b) $\{\ldots., a^{-2}, a^{-1}, a^{-0} = e, a^1, a^2, \ldots\}$ is a cyclic group with generator a.

(c) Under multiplication $\{1, -1, i, -i\}$ is a cyclic group with i as a generator, $-i$, being a generator also.

5.1.3 *Theorem*

1. Every cyclic group is abelian.
2. The order of a cyclic group is same as that of its generator.

Proof

1. Let $G = (a) = \{e, a, a^2, a^3, \ldots\}$ be a cyclic group with a as the generator. Then for any $m, n \in Z, a^m, a^n \in G$ and $a^m a^n = a^{m+n}$ $= a^{n+m} = a^n a^m$ and hence G is abelian. In the additive notation we write,

   $$G = \{1, a, 2a, 3a, \ldots\}$$

In this case

$ma + na = (m + n)a = (n + m)a = na + ma$

and thus G is abelian.

2. Let G be a cyclic group with generator $a \in G$. Let $o(a) = n$, i.e.

   $a^n = e$ and $a^m \neq e$ for $0 < m < n$

   if $m > n$, let $m = qn + r, 0 \leq r \leq n$.

   So

   $a^m = a^{qn+r} = a^{qn}.a^r = (a^n)^q.a^r$

   $= e^q.a^r = e.a^r = a^r$

   Therefore there are exactly n elements in the group and they are:

   $\{e, a, a^2, \dots, a^{n-1}\}$

   This completes the proof.

5.2 SUBGROUPS

5.2.1 *Definition*

Let G be a group and let H be a nonempty subset of G. H is said to be a subgroup of G if under the composition in G, H itself forms a group. If H is a subgroup of G then we denote this by $H < G$.

5.2.1 *Examples*

1. The singleton set $\{e\}$ and the whole group G, where e is the identity element of G are subgroups of any group G. These are called trivial subgroups or improper subgroups. Any other subgroup is called a proper or nontrivial subgroup.
2. The group of all even integers is a subgroup of the group of all integers under addition. But the set of all odd integers is not a subgroup of all integers under addition. Similarly the set of all positive integers is not a subgroup of all integers under addition.
3. The group of all integers is a subgroup of all rational numbers under addition.
4. The group of positive rational numbers under multiplication is a subgroup of nonzero real numbers under multiplication.
5. $\{1, -1, i, -i\}$ is a subgroup of the group of all nonzero complex numbers under multiplication.
6. $\{1, -1\}$ is a subgroup of $\{1, -1, i, -i\}$ under multiplication.

5.2.1 *Theorem*

Let G be a group and let $\phi \neq H \subset G$.

Then $H < G$ if and only if:

(i) $a, b \in H$ if implies that $ab \in H$.

(ii) $a, b \in H$ implies that $a^{-1} \in H$.

Proof: If $H < G$ then it clearly follows from the definition that (i) and (ii) hold.

Conversely, suppose that $\phi \neq H \subset G$ for which (i) and (ii) hold. Since associative law holds in G and $H \subset G$ it also holds in H. So in order to prove that $H < G$ it is now enough to show that $e \in H$. For this, let $a \in H$. By (ii) it follows that $a^{-1} \in H$. Now by (i) we get $e = aa^{-1} \in H$ and this completes the proof.

5.2.2 *Theorem*

Let G be a group and let $\phi \neq H \subset G$. Then $H < G$ if and only if $ab^{-1} \in H$ whenever $ab \in H$.

Proof: Let $H < G$ and let $a, b \in H$. Since $b \in H$ we have $b^{-1} \in H$. Again since $a \in H$ we have $ab^{-1} \in H$.

Conversely, let $\phi \neq H \subset G$ for which $ab^{-1} \in H$ whenever $a, b \in H$. Let $a \in H$. Then $aa^{-1} = e \in H$. Now for each $b \in H$ we have $eb^{-1} = b^{-1} \in H$. Let $a, b \in H$. Then $b^{-1} \in H$ and hence $a(b^{-1})^{-1} = ab \in H$. The associative law clearly holds and hence $H < G$.

5.2.3 *Theorem*

If H is a finite nonempty subset of a group G and H is closed under the composition, then H is a subgroup of G.

Proof: We need only to show that whenever $a \in H$ then $a^{-1} \in H$. Let $a \in H$. Then since H is closed under the composition we have $aa = a^2 \in H, a^2a = a^3 \in H, a^3a = a^4 \in H, \dots, a^m \in H, \dots$ Thus the infinite set of elements $a, a^2, a^3, \dots, a^m, \dots$ must lie in H. But since H is a finite set, there must be some repetitions in the above set of elements, i.e. for some integers r and s with $r > s > 0$ we must have $a^r = a^s$. Therefore $a^{r-s} = a^{s-s} = e \in H$. Since $r - s - 1 \geq 0, a^{r-s-1} \in H$. Further since $aa^{r-s-1} = a^{r-s} = e$, we have $a^{-1} = a^{r-s-1}$. Thus $a^{-1} \in H$. This completes the proof of the theorem.

5.2.4 *Theorem*

If H_1 and H_2 are subgroups of a group G then $H_1 \cap H_2 < G$.

Proof: Since H_1 and H_2 are subgroups of a group $G, e \in H_1 \cap H_2$ and thus $H_1 \cap H_2 \neq \varphi$. Let $a, b \in H_1 \cap H_2$. Then $a, b \in H_1$ and since $H_1 < G. ab^{-1} \in H_1$. Similarly $ab^{-1} \in H_2$. Thus $ab^{-1} \in H_1 \cap H_2$ and shows that $H_1 \cap H_2 < G$.

5.2.5 *Theorem*

If H_1 and H_2 are subgroups of a group G, then $H_1 \cup H_2 < G$ if and only if,

$$H_1 \subset H_2 \; or \; H_2 \subset H_1.$$

Proof: If H_1 and H_2 are subgroups of a group G and if either $H_1 \subset H_2$ or $H_2 \subset H_1$ then it is clear that, $H_1 \cup H_2 < G$.

Conversely, let $H_1 \cup H_2 < G$. We have to show that either $H_1 \subset H_2$ or $H_2 \subset H_1$. Assume, to the contrary, that neither holds. Then it is possible to choose elements a and b such that $a \in H_1 - H_2$ and $b \in H_2 - H_1$.

In other words $a \in H_1, a \notin H_2, b \in H_2, b \notin H_1$ but $a, b \in H_1 \cup H_2$. Since $H_1 \cup H_2$ is a subgroup and $a, b \in H_1 \cup H_2$ we must have $ab \in H_1 \cup H_2$. Therefore either $ab \in H_1$ or $ab \in H_2$. Let $ab \in H_1$. Since $a \in H_1$ we have $a^{-1} \in H_1$ and

$$b = a^{-1}(ab) \in H_1$$

which contradicts the fact that $b \notin H_1$. On the other hand if $ab \in H_2$. Since $b \in H_2$ and thus $b^{-1} \in H_2$ we must have

$$a = (ab)b^{-1} \in H_2$$

But this is impossible. This contradiction completes the proof.

5.2.2 *Definition*

If H and K are any two subsets of a group G then we define,

$$HK = \{x \in G: x = hk, h \in H, k \in K\}$$

In the following theorem we shall assume that H and K are any two subgroups of a group G.

5.2.6 *Theorem*

HK is a subgroup of G if and only if $HK = KH$. In particular if G is abelian, then HK is a subgroup of G.

Proof: Suppose that $HK = KH$. In other words if $h \in H$ *and* $k \in K$ then $hk = k_1h_1$ for some $k_1 \in K$ and $h_1 \in H$. We shall show that HK is a subgroup of G. First let $x = hk \in HK$ and $y = h'k' \in HK$. Then $xy = hkh'k'$. Now since $kh' \in KH = HK$ we have $kh' = h_2k_2$ for some $h_2 \in H$ and $k_2 \in K$. Hence $xy = h(kh')k' = h(h_2k_2)k' = (hh_2)(k_2k') \in HK$. Further $x^{-1} = (hk)^{-1}h^{-1} \in KH = HK$ and thus $x^{-1} \in HK$ for each $x \in HK$. Therefore HK is a subgroup of G.

Conversely, let HK be a subgroup of G. Then for any $h \in H$ and $k \in K, h^{-1} \in H. k^{-1} \in K$ and so $h^{-1}k^{-1} \in HK$. Therefore $kh = (h^{-1}k^{-1})^{-1} \in HK$.

Thus $KH \subset HK$. Now for any element $x \in HK, x^{-1} = hk \in HK$ so $x = (x^{-1})^{-1} = hk^{-1} = k^{-1}h^{-1} \in KH$ showing that $HK \subset KH$ and this completes the proof. The particular case follows from the fact that if G is abelian then $HK = KH$.

If H and K are finite subgroups of a group G, then although HK need not be a sub-group, it is clearly a finite set. Therefore it is a natural and a meaningful question to ask: How many distinct elements are there in the subset HK. If $o(HK)$ denotes the number of distinct elements in the set HK then we have.

5.2.7 *Theorem*

If H and K are finite subgroups of a group G of orders $o(H)$ and $o(K)$ respectively, then

$$o(HK) = \frac{o(H)o(K)}{o(H \cap K)}$$

In particular if $H \cap K = \{e\}$ *then* $o(HK) = o(H)o(K)$.

Proof: Let $h \in H$ and $k \in K$ so that $hk \in HK$. If $h_1 \in H \cap K$ then $hk = (hh_1)\,({h_1}^{-1}k)$ where $hh_1 \in H$ since $h \in H$ and $h_1 \in H \cap K \subset H$ and ${h_1}^{-1} k \in K$ since ${h_1}^{-1} \in H \cap K \subset K$ and $k \in K$. Thus the element hk is duplicated atleast $o(H \cap K)$ times. However, if $hk = h_2k_2$, then $h^{-1}h_2 = k(k_2)^{-1} = u$ and $u \in H \cap K$ and so $h_2 = hu, k_2 = u^{-1}k$ thus all the duplications are in the form $hk = (hh_1)({h_1}^{-1}k)$. Therefore the element hk appears in the list of HK exactly $o(H \cap K)$ times. But the total number in the listing of HK is $o(H)o(K)$. Therefore the number of distinct elements in $HK, o(HK)$ is given,

$$\frac{o(H)o(K)}{o(H \cap K)}$$

We shall now introduce two subgroups of a group G which are very useful and important.

5.2.3 *Definition*

If G is a group and $a \in G$ the normalizer of a in G is the set,

$$N(a) = \{x \in G: xa = ax\}$$

5.2.8 *Theorem*

$N(a)$ is a subgroup of G.

Proof: Suppose that $x, y \in N(a)$. Then $xa = ax$ and $ya = ay$. Therefore $(xy)a = x(ya) = x(ay) = (xa)y = (ax)y = a(xy)$ and so $xy \in N(a)$.

Now since $ax = xa$ it follows that,

$$x^{-1}a = x^{-1}(ax)x^{-1} = x^{-1}(xa)x^{-1} = ax^{-1}$$

So that $x^{-1} \in N(a)$ and this completes the proof.

5.2.4 *Definition*

If G is a group then the center of G is defined by,

$$cent\ G = \{c \in G: xc = cx \text{ for all } x \in G\}$$

5.2.9 *Theorem*

Cent G is a subgroup of G.

Proof: First of all observe that cent $G \neq \varphi$ since $e \in cent\ G$. Let $a, b \in cent\ G$. Then $xa = ax$ and $xb = bx$ for all $x \in G$. Therefore $b^{-1}(xb)b^{-1} = b^{-1}(bx)b^{-1}$.

Or

$$b^{-1}x = xb^{-1}$$

Now for any $x \in G, (ab^{-1})x = a(b^{-1}x) = a(xb^{-1}) = (ax)b^{-1} = (xa)b^{-1} = x(ab^{-1})$.

Therefore $ab^{-1} \in cent\ G$ and this proves that $cent\ G$ is a subgroup of G.

5.2.5 *Definition*

Let G be a group and H be a subgroup of G. Let $a, b \in G$. a is said to be congruent to $b\ mod\ H$ written as $a \equiv b\ (mod\ H)$ if $ab^{-1} \in H$.

5.2.10 *Theorem*

The relation $a \equiv b\ (mod\ H)$ is an equivalence relation.

Proof: We have to show that for $a, b, c \in G$

1. $a \equiv a\ (mod\ H)$ (reflexivity)
2. $a \equiv b\ (mod\ H)$implies $b \equiv a\ (mod\ H)$ (symmetry)
3. $a \equiv b\ (mod\ H), b \equiv c\ (mod\ H)$ implies $a \equiv c\ (mod\ H)$ (transitivity).

Since H is a subgroup we have $e \in H$. But for any $a \in G, e = aa^{-1}$. Hence $aa^{-1} \in H$ and this proves the reflexivity. For the symmetry, let $a \equiv b\ (mod\ H)$. Then $ab^{-1} \in H$. Since H is a subgroup.

$$(ab^{-1})^{-1} = (b^{-1})^{-1}a^{-1} = ba^{-1} \in H$$

Therefore $b \equiv a\ (mod\ H)$ and this establishes (ii). Now let $a \equiv b\ (mod\ H)$ and $b \equiv c\ (mod\ H)$.

Then $ab^{-1} \in H$ and $bc^{-1} \in H$. Now,

$$ac^{-1} = a(b^{-1}b)c^{-1} = (ab^{-1})(bc^{-1}) \in H$$

Hence $a \equiv c\ (mod\ H)$ and this completes the proof.

5.2.2 *Example*

Let G be the group Z of all integers under addition. Let n be a fixed but arbitrary integer and let $H = H_n$ be the subgroup of Z consisting of all multiples of n. Then for $a, b \in G$ the relation $a \equiv b\ (mod\ H)$, i.e, $ab^{-1} \in H$ in this particular case under additive notation reads $a \equiv b\ (mod\ n)$ if $a - b$ is a multiple of n.

This is the usual congruence $mod\ n$ discussed in Number Theory. Thus the relation defined above is the natural generalization of a familiar group to any arbitrary group and its subgroup.

We shall now elaborate the usual congruence relation by example.

Let n be a positive integer and $n|(x - a)$. Then x is congruent to $a\ mod\ n$ and written as $x \equiv a\ (mod\ n)$. On the other hand if $n \nmid (x - a)$, then x is said to be incongruent $a\ (mod\ n)$ and written as $x \not\equiv a\ (mod\ n)$.

Let $x \equiv a\ (mod\ n)$. Then a is said to be a residue of $x\ mod\ n$. Again if $0 = \leq a \leq n - 1$ then a is said to be the least residue of $x\ mod\ n$.

For a given residue a the set of all integers x congruent to $a (mod\ n)$ is called the class of residues $mod\ n$. The difference of any two elements of this class is divisible by n.

Let $n = 4, a = 3$. Then if $x \equiv 3\ (mod\ 4), 4|(x-3), i.e. x - 3 = 4q, x = 4q + 3$. By taking different integer values of q we obtain the set of values of x as $\{..., -17, -13, -9, -5, -1, 3, 7, 11, 15, 19, ...\}$

We denote this set by $\overline{3}$. Similarly,

$$\overline{0} = \{..., -4, 0, 4, 8, ...\}$$
$$\overline{1} = \{..., -7, -3, 1, 5, 9, ...\}$$
$$\overline{2} = \{..., -6, -2, 2, 6, 10, ...\}$$

Similarly, if $m = 5$ we get

$$\overline{0} = \{..., -15, 10, -5, 0, 5, 10, ...\}$$
$$\overline{1} = \{..., -14, -9, -4, 1, 6, 11, ...\}$$
$$\overline{2} = \{..., -13, -8, -3, 2, 7, 12, ...\}$$
$$\overline{3} = \{..., -12, -7, -2, 3, 8, 13, ...\}$$
$$\overline{4} = \{..., -11, -1, -3, 4, 9, 14, ...\}$$

It is clear that for a given n, the n number of residue classes are as follows:

$$\overline{0} = \{..., -3n, -2n, -n, 0, n, 2n, 3n, ...\}$$
$$\overline{1} = \{..., -3n + 1, -2n + 1, -n + 1, 1, n + 1, 2n + 1, 3n + 1, ...\}$$
$$\overline{2} = \{..., -2n + 2, -2n + 2, -n + 2, 2, n + 2, 2n + 2, 3n + 2, ...\}$$

....

$$\overline{r} = \{..., -3n + r, -2n + r, -n + r, r, n + r, 2n + r, 3n + r, ...\}$$
$$\overline{n-1} = \{..., -2n - 1, -n - 1, -1, n - 1, 2n - 1, 3n - 1, ...\}$$

5.2.6 *Definition*

Let H be a subgroup of a group G and let $a \in G$. Then the set,

$$Ha = \{ha: h \in H\}$$

is called a right coset of H in G. Left cosets are defined similarly. Note that if G is an abelian group then the left cosets and right cosets are identical.

5.2.11 *Theorem*

For all $a \in G, Ha = \{x: x \in G$ and $a \equiv x\ mod\ H\}$

Proof: For $a \in G, Ha = \{ha: h \in H\}$

Let $[a] = \{x: a \equiv x\ mod\ H\}$. We shall first show that $Ha \subset [a]$. For $h \in H, a(ha)^{-1} = a(a^{-1}h^{-1}) = h^{-1}$. So $a(ha)^{-1} \in H$, i.e. $a \equiv (ha)\ mod\ H$. So $ha \in [a]$. Hence for every $h \in H$ we have $Ha \subset [a]$.

Now suppose that $x \in [a]$

Then $ax^{-1} \in H$ and so $(ax^{-1})^{-1} = xa^{-1} \in H, i.e. xa^{-1} = h$ for some $h \in H$. Multiplying both sides by a on the right, we get $x = ha$ so $x \in Ha$. Hence $[a] \subset Ha$. Thus we have $[a] = Ha$.

Remark

These equivalence classes yield a decomposition of G into disjoint classes. Any two right cosets of H in G are either identical or have no element in common. We have the following theorem.

5.2.12 *Theorem*

There is a 1-1 correspondence between any two right cosets of H in G.

Proof: Let Ha and Hb be any two right cosets of H in G.

Define mapping $f: Ha \to Hb$ by

$$f(ha) = hb, h \in H$$

Now f is 1-1, for if $f(h_1 a) = f(h_2 a)$ then $h_1 b = h_2 b$ and so $h_1 = h_2$. f is clearly onto, for $hb \in Hb \Rightarrow ha \in Ha$ and $f(ha) = hb$.

Remark

If H is a subgroup of a finite group G then it follows from the above theorem that any two right cosets of H in G have the same number of elements. $H = He$ is itself a right coset of H in G. $o(H)$ denotes the order of H. It now follows that $o(H) = o(He)$ and hence the number of elements in a right coset of H in G is equal to $o(H)$. We have the following theorem.

5.2.13 *Theorem (Lagrange's Theorem)*

If G is finite group and H is a subgroup of G then $o(H)$ is a divisor of order of G.

Proof: Let G be a finite group and H a subgroup. Let k be the number of distinct right cosets of H in G.

Any two distinct right cosets have no element in common and the number of elements in each is equal to $o(H)$.

Therefore $k\, o(H) = o(G)$ and this completes the proof.

5.2.7 *Definition*

Let H be a subgroup of G. Index of H in G is the number of distinct right cosets of H in G and is denoted by $i_G(H)$. If G is a finite group then by Lagrange's theorem we have,

$$i_G(H) = \frac{o(G)}{o(H)}$$

Corollary 1

If G is a finite group and $a \in G$, then $o(a)|o(G)$

Proof: Consider the subgroup $\{e, a, a_2, a_3, ...\} = (a)$

We claim that the number of $(a) = o(a)$. Since $a^{o(a)} = e$ the subgroup has atmost $o(a)$ elements. If the number of elements is less than $o(a)$, then $a^i = a^j$ for some integers $0 \leq i < j < o(a)$.

So $a^{j-i} = e$ when $0 < j - i < o(a)$ and this is a contradiction.

Thus the cyclic subgroup generated by a has $o(a)$ elements. By Lagrange's theorem $o(a)|o(G)$.

Corollary 2

If G is a finite group and $a \in G$ then $a^{o(G)} = e$.

Proof: We have $o(a)|o(G)$. So $o(G) = mo(a)$ for some $m \in N$. Therefore $a^{o(G)} = (a^{o(a)})^m = e^m = e$.

Corollary 3

If G is a finite group whose order is a prime number p then G is a cyclic group.

Proof: We claim that G has no nontrivial subgroup H. For, $o(H)$ must divide $o(G) = p$ so that:

$$o(H) = p \text{ or } o(H) = 1$$

If $o(H) = 1$ then $H = \{e\}$
If $o(H) = p$ then $H = G$

Suppose that $a \neq e \in H$ and $H = (a)$, H is a subgroup of G, $H \neq \{e\}$ since $a \neq e \in H$. Thus $H = G$. From this it follows that G is cyclic and that every element of G is a power of a.

5.2.3 *Example*

1. If H is a subgroup of all even integers of the group Z of all integers under addition, then cosets of H are as follows:

   $H = H + 0 = \{\dots, -6, -4, -2, 0, 2, 4, 6, \dots\}$

   $H + 1 = \{\dots, -5, -3, -1, 1, 3, 5, 7, \dots\}$

   $H + 2 = \{\dots, -4, -2, 0, 2, 4, 6, \dots\}$

   $H + 3 = \{\dots, -3, -1, 1, 3, 5, \dots\}$

   $H + 4 = \{\dots, -2, 0, 2, 4, \dots\}$

   $H + 5 = \{\dots, -1, 1, 3, 5, 7, \dots\}$

   It should be noted that

   $H = H + 2 = H + 4 = \cdots$

   $H + 1 = H + 3 = H + 5 = \cdots$

   Therefore two distinct cosets are H *and* $H + 1$ and $Z = H \cup (H + 1)$. It should also be noted here that since Z is an abelian group under addition the left cosets and the right cosets of H are identical.

2. Find out the subgroups and their order of the group of all permutations of the three elements set $\{a, b, c\}$.

   This set has 3!=6 permutations which are as follows:

   $$I = P_0 = \begin{pmatrix} a & b & c \\ a & b & c \end{pmatrix}, P_1 = \begin{pmatrix} a & b & c \\ a & c & b \end{pmatrix}, P_2 = \begin{pmatrix} a & b & c \\ b & a & c \end{pmatrix}$$

   $$P_3 = \begin{pmatrix} a & b & c \\ b & c & a \end{pmatrix}, P_4 = \begin{pmatrix} a & b & c \\ c & a & b \end{pmatrix}, P_5 = \begin{pmatrix} a & b & c \\ c & b & a \end{pmatrix}$$

   Now $\{P_0, P_1, P_2, P_3, P_4, P_5\}$ is a group under multiplication of permutations and its order is 6.

   The sets $\{P_0, P_1\}$, $\{P_0, P_2\}$ and $\{P_0, P_5\}$ are subgroups of the above group each of order 2. The set $\{P_0, P_3, P_4\}$ is a subgroup of order 3. The group itself is a subgroup of order 6 and $\{P_0\}$ is a subgroup of order 1. The group does not have any subgroup of order 4 or 5 (see Lagrange's theorem). All the subgroups are of order 1, 2, 3, and 6, each being a divisor of 6 (the order of the group).

2.5.8 *Definition*

Let G be a group and H a subgroup of $G.H$ is said to be a normal subgroup of G if every left coset of H in G is a right coset of H in G. If H is normal subgroup of G it is denoted by $H\Delta G$. In other words, if for each $a \in G$ we have $Ha = aH$ then $H\Delta G$.

Remark

1. Note that $Ha = aH$ does not mean that $ha = ah$ for every $h \in H$. It really means that for every $h \in H$ there exists some $h_1 \in H$ such that $ha = ah_1$.
2. The trivial subgroups $\{e\}$ and G are always normal subgroups.
3. A group may not have any nontrivial normal subgroup. Such groups are called simple groups.

   We now give an equivalent characterization of a normal subgroup. If H is a subgroup of a group G and if $a \in G$ then we write,

   $aHa^{-1} = \{aha^{-1} : h \in H\}$

5.2.14 *Theorem*

1. H is a normal subgroup of G if and only if $aHa^{-1} \subset H$ for each $a \in G$.
2. Moreover $H \Delta G$ if and only if $aHa^{-1} = H$ for each $a \in G$.

Proof: Assume that $H\Delta G$. Therefore for each $a \in G$ we have $aH = Ha$. Let $h_1 \in H$, i.e., $ah_1a^{-1} \in aHa^{-1}$. Since $aH = Ha$ we can choose $h_2 \in H$ such that $ah_1 = h_2a$, so that $ah_1a^{-1} = (h_2a)a^{-1} = h_2 \in H$. Therefore $aHa^{-1} \subset H$. Further,

$$a^{-1}Ha = a^{-1}H(a^{-1})^{-1} \subset H.$$

Now since $a^{-1}Ha \subset H, H = a(a^{-1}Ha)a^{-1} \subset aHa^{-1} \subset H$ and hence $aHa^{-1} = H$.

Conversely assume that for each $a \in G$, $aHa^{-1} \subset H$. We have to show that $H\Delta G$, in other words $aH = Ha$. Now for any $h \in H, ah \in aH$. Since $aHa^{-1} \subset H, aha^{-1} \in H$. So we can choose some $h_1 \in H$ such that $aha^{-1} = h_1, i.e\ ah = h_1a$. Since $h_1a \in Ha, ah \in Ha$.

Since $a^{-1} \in G$ we have,

$$a^{-1}Ha = a^{-1}H(a^{-1})^{-1} \subset H$$

So we can choose h and h_1 in such a manner that,

$$a^{-1}h(a^{-1})^{-1} = h_1\ or\ a^{-1}ha = h_1\ or\ ha = ah_1$$

Since $ha \in Ha$ so $ha \in aH$. Thus $Ha \subset aH$ and hence $Ha = aH$. Now if $aHa^{-1} = H$ for every $a \in G$ certainly $aHa^{-1} \subset H$ and therefore the result follows.

5.2.4 *Example*

1. If G is a group then cent G, the centre of G, is a normal subgroup of G. We have already demonstrated.
   That cent G is a subgroup of G. Note that if $c \in Cent\ G$ then from the definition of Cent G, it follows that $ca = ac$. Therefore aca^{-1} $= (ca)a^{-1} = c(aa^{-1}) = c \in cent\ G$. Hence $cent\ G \Delta G$.
2. If H_1 and H_2 are normal subgroups of G then $H_1 \cap H_2$ is a normal subgroup of G. We have proved earlier that if H_1 and H_2 are subgroup of G then $H_1 \cap H_2$ is also a subgroup of G. Now let $h \in H_1 \cap H_2$. Then $h \in H_1$ and $h \in H_2$. Since H_1 and H_2 are normal subgroups, for each $a \in G, aha^{-1} \in H_1$ and $aha^{-1} \in H_2$. Therefore $aha^{-1} \in H_1 \cap H_2$ and thus $H_1 \cap H_2 \Delta G$.
3. Every subgroup of an abelian group is normal. To show this, let G be an abelian group and let $H < G$. We have to show that if $h \in H$ then for every $a \in G, aha^{-1} \in H$. Let $h \in H$. Then for any $a \in G\ aha^{-1} =$ $(ah)a^{-1} = (ha)a^{-1}$ (since G is abelian), $= h(aa^{-1}) = he = h \in H$. However, the converse need not be true, in other words, if every subgroup of a group G is normal then G need not be abelian (Give an example).

5.2.15 *Theorem*

A subgroup H of G is a normal subgroup of G if and only if the Product of two right cosets of H in G is again a right coset of H in G.

Proof: Let $H \Delta G$ and let $a, b \in G$. First of all note that for any subgroup H of G, $HH = H$. Now consider the product $(Ha)\,(Hb)$ of two right cosets Ha and Hb of $H\ in\ G$. Since $H\ \Delta\ G, aH = Ha$ and so

$$(Ha)\,(Hb) = H(aH)b = H(Ha)b = HHab = Hab$$

Thus the product of two right cosets is again a right coset.

Conversely, let $H < G$ and let the product of any two right cosets of $H\ in\ G$ be a right coset of $H\ in\ G$.

For $a \in G$, Ha and Ha^{-1} are two right cosets of $H\ in\ G$ and hence $HaHH^{-1}$ is also a right coset of $H\ in\ G$. Now $eaea^{-1} \in HaHa^{-1}$, but $eaea^{-1} =$ $aa^{-1} = e \in H$ and H is itself a right coset. Further if two right cosets have one element common, they must be identical, Hence for all $a \in G, H =$ $HaHa^{-1}$. Let $h \in H$ and $a \in G$. Then

$$aha^{-1} = (ea)(ha^{-1}) \in (Ha)(Ha^{-1}) = H$$

and this completes the proof.

We have observed earlier that if H and K are subgroups of a group G then HK need not be a subgroup, it is, however, a subgroup if and only if $HK = KH$ or in particular if G is abelian. We shall now show that if one of them is a normal subgroup, then the product is a subgroup and if both are normal, the product is a normal subgroup.

5.2.16 *Theorem*

Let G be a group, $N \Delta G$ and $H < G$. Then $NH < G$. If further $H \Delta G$, then $NH \Delta G$.

Proof: Let $nh \in NH$ and $n_1h_1 \in NH$. Then

$$(nh)\,(n_1h_1)^{-1} = (nh)\,({h_1}^{-1}{n_1}^{-1})$$
$$= n(h{h_1}^{-1}){n_1}^{-1} = nh_2{n_1}^{-1} \text{ (say } h{h_1}^{-1} = h_2)$$

Since $H < G$, $h{h_1}^{-1} = h_2 \in H$. Also, $h_1 \in H$. Since $N\Delta G, Na = aN$ for each $a \in G$. Now $h_2 \in H \subset G$, so $Nh_2 = h_2N$. Again if $n_1 \in N, {n_1}^{-1} \in N$. We can now choose $n_2 \in N$ such that $n_2h_2 = h_2{n_1}^{-1}$.

Therefore,

$$(nh)\,(n_1h_1)^{-1} = n(h_2n_1)^{-1}$$
$$= n(n_2h_2)$$
$$= (nn_2)h_2 - n_3h_2 \in NH$$

Since $nn_2 = n_3 \in N$. Therefore it follows that $NH \subset G$.

We shall now show that NH is normal in G. Let $nh \in NH$. Then for any $a \in G$

$$a(nh)a^{-1} = (an)\,(ha^{-1})$$
$$= \big((an)e\big)(ha^{-1})$$
$$= (an)(a^{-1}a)(ha^{-1})$$
$$= (ana^{-1})(aha^{-1}) \in NH$$

Since N and H are normal subgroups, $ana^{-1} \in N$ and $aha^{-1} \in H$. Thus $NH \Delta G$ and this completes the proof.

We have already proved that the intersection of any two subgroups of a group G is again a subgroup. We shall now show that if one of them is a normal subgroup, the intersection is a normal subgroup.

5.2.17 *Theorem*

Let G be a group, $H < G$ and $N \Delta G$. Then $H \cap N \Delta H$.

Proof: Since $H \cap N \subset H$ and $H \cap N < G$. So that $H \cap N < H$. Let $a \in H$ and $h \in H \cap N$. Then $h \in H$ and $h \in N$. Since $N \Delta G$, $aha^{-1} \in N$. Since $h \in H, a \in H$ we have $ah \in H$.

Further $a^{-1} \in H$, so $(ah)a^{-1} \in H$. Therefore $(aha^{-1}) \in H \cap N$ and this shows that $H \cap N$ is a normal subgroup of H.

We have already observed that for a normal subgroup of a group G the product of two right cosets is again a right coset. We shall now demonstrate that we can use this product to make the collection of right cosets into a group, which will be called the quotient group or the factor group. This type of structures which are called quotient structures are very important and useful in mathematics.

If H is a normal subgroup in a group G, G/H denotes the collection of all right cosets of H in G, i.e.

$$G/H = \{Ha: a \in G\}$$

Recall that for $X = Ha \in G/H$ and $Y = Hb \in G/H$ for some $a, b \in G$ we define,

$$XY = NaNb = Nab \in G/H$$

We have.

5.2.18 *Theorem*

If G is a group and $N \Delta G$, then G/H is a group (called the quotient group or factor group) under the product defined as above.

Proof: Let $X = Na, Y = Nb$ and $Z = Nc$ be in $\frac{G}{H}$ with $a, b, c \in G$

Then,

$$(XY)Z = (NaNb)Nc = NabNc = N(ab)c = Na(bc)$$
$$= NaNbc = Na(NbNc) = X(YZ)$$

and hence associativity holds. Now $H = He \in G/H$ is the identity element of G/H, for if $X = Ha \in G/H$ for some $a \in G$, then

$$XH = HaHe = Hae = Ha = X$$

and similarly $HX = X$.

Suppose $X = Ha \in G/H$ where $a \in G$.

Then $Ha^{-1} \in G/H$ and is the inverse of Ha, since $HaHa^{-1} = Haa^{-1} = He = H$, and similarly $Ha^{-1}Ha = H$.

This completes the proof of the theorem.

Remark

(a) If G is a finite group and H is a normal subgroup of G, then $o(G/H)$ = the number of elements in G/H = the number of distinct right cosets of H in $G = i_G(H)$ = the index of H in G. Therefore by using Lagrange's theorem we have,

$$o(G/H) = \frac{o(G)}{o(H)}$$

(b) If G is abelian and $H < G$, then the quotient group $\frac{G}{H}$ is also abelian. To prove this let G be abelian and let $H < G$. Then H is also abelian and $H \Delta G$. Let $X = Ha, Y = Hb$ be in $\frac{G}{H}$ for $a, b \in G$. Then $XY = HaHb = Hab = Hba$ (since G is abelian) $= HaHb = YX$.

And this proves the result.

5.3 HOMOMORPHISM

In this section we shall discuss about the mappings from one group into another which preserve the algebraic operations. Such mappings are called homomorphisms and if there is one central idea which is common to all aspects of algebra, it is the concept of homomorphism. The ideas and results here are closely connected with and use the concept of normal subgroups.

Definition

Let G and $\bar{G}$ be groups and let $f: G \rightarrow \bar{G}$. f is said to be a homomorphism from G into $\bar{G}$ if for all $a, b \in G, f(ab) = f(a)f(b)$.

A homomorphism f from G into $\bar{G}$ is called an isomorphism if f is 1–1 and two groups G and $\bar{G}$ are said to be isomorphic if there exists an isomorphism from G onto $\bar{G}$. A homomorphism of a group G into itself is called an endomorphism. An isomorphism of a group G into itself is called an automorphism. If f is a homomorphism from G into $\bar{G}$ then the kernel of f, denoted by K_f, is defined by,

$$Kf = \{x \in G: f(x) = \bar{e} = \text{the identity element of } \bar{G}\}.$$

Remark

Note that in the definition of homomorphism for $a, b \in G$, by ab we mean the composition in G and by $f(a)\,f(b)$ we mean the composition in $\bar{G}$. For the convenience we use the same symbol and it should not create any confusion.

5.3.1 *Example*

(a) Note that the mapping $f: G \to \bar{G}$ defined by $f(x) = \bar{e}$ for all $x \in G$ is trivially a homomorphism (in fact an endomorphism) with $K_f = G$.

(b) The map $f: G \to \bar{G}$ defined by $f(x) = x$ for all $x \in G$ is a homomorphism (an automorphism) with $K_f = \{e\}$.

(c) Let G be the group C of all complex numbers under addition. Then the map $f: C \to C$ defined by $f(z) = \bar{z}$ for all $z \in C$ is an automorphism.

(d) Let $G = Z$, the group of integers under addition. Then $f: G \to G$ defined by $f(x) = 2x$ is a homomorphism.

(e) Let G = the group of real numbers under addition and $\bar{G}$ = the group of nonzero real numbers under multiplication. Then $f: G \to \bar{G}$ defined by $f(a) = 2^a$ is a homomorphism.

(f) Let G be a group and $H < G$. Then the inclusion map $i: H \to G$ defined by $i(a) = a$ for all $a \in H$ is a homomorphism from H into G.

(g) Let G = the group of nonzero reals under multiplication and $\bar{G} = \{1, -1\}$. Then $f: G \to \bar{G}$ defined by

$$f(x) = \begin{cases} 1, x \text{ positive} \\ -1, x \text{ negative} \end{cases}$$

is a homomorphism.

(h) Let G = the group of integers under addition and $\bar{G}$ = the group of integers under addition modulo n. Define f by $f(x)$ = remainder of x on division by n. Then f is a homomorphism.

(i) G = the group of positive real numbers under multiplication and $\bar{G}$ = the group of all real numbers under addition. Define $f: G \to \bar{G}$ by $f(x) = \log_{10} x$. Then f is a homomorphism and it is also 1–1 and onto.

(j) Let G be the group of all 2×2 matrices $\begin{pmatrix} a & b \\ c & d \end{pmatrix}$ such that $ad - bc \neq 0$, under matrix multiplication. Let $\bar{G}$ be the group of all nonzero real numbers under multiplication. Define $f : G \to \bar{G}$ by

$$f\left(\begin{pmatrix} a & b \\ c & d \end{pmatrix}\right) = ad - bc$$

then f is a homomorphism from G onto $\bar{G}$.

(k) Let $G =$ the group of all integers under addition and $\bar{G}$ = the group of positive real numbers under, multiplication. Then $f: G \to \bar{G}$ defined by $f(x) = e^x$ is an isomorphism.

(l) Any finite cyclic group G of order n is isomorphic to the group of residue class mod n under addition mod n.

If G is a finite cyclic group of order n with generator a and $\left\{\bar{0}, \bar{1}, \ldots, \overline{n-1}\right\}$ is the group of residue class mod n, then the map given by $f(a') = \bar{r}$ for $r = 0, 1, \ldots, n-1$, defines an isomorphism. For example, $G = \{1, -1\}$ under multiplication is isomorphic to $\bar{G} = \{0, 1, 2, 3 \text{ (mod) } 4\}$ under addition.

(m) Any infinite cyclic group G is isomorphic to Z, the group of integers under addition. If a is a generator of G then the map f given by,

$f(a^k) = k$

is an isomorphism.

(n) The group Z of integers under addition is not isomorphic to the group of nonzero rational numbers under multiplication. (Prove this!)

5.3.1 *Theorem*

Let f be a homomorphism from G into $\bar{G}$. Then

(a) $f(e) = \bar{e}$ where $\bar{e}$ is the unit element of $\bar{G}$,

(b) $f(a^{-1}) = f(a)^{-1}$ for each $a \in G$.

Proof

(a) For each $a \in G$, $f(a)\bar{e} = f(a) = f(ae) = f(a)f(e)$ and hence $f(e) = \bar{e}$.

(b) For each $a \in G, f(a)f(a^{-1}) = f(aa^{-1}) = f(e) = e = f(a)f(a)^{-1}$ and hence $f(a^{-1}) = f(a)^{-1}$.

We shall now show that the kernel of a homomorphism has very interesting property, namely it is a normal subgroup.

5.3.2 *Theorem*

Let f be a homomorphism from G into $\bar{G}$ with the kernel K_f, Then K_f is a normal subgroup of G.

Proof: First of all note that $K_f \neq \varphi$ since $e \in K_f$, we now check that $K_f \in G$. For this purpose, let $a, b \in K_f$. Then $f(a) = \bar{e}$ and $f(b) = \bar{e}$. Now

$f(ab^{-1}) = f(a)f(b^{-1}) = \bar{e}f(b)^{-1} = \overline{ee}^{-1} = \overline{ee} = e$. Hence $ab^{-1} \in K_f$, and thus $K_f < G$. For the normality of K_f, we have to show that for every $a \in G$ and $k \in K_f$, $aka^{-1} \in K_f$. Now let $a \in G$ and $k \in K_f$. Then

$$f(aka^{-1}) = f(a)f(k)f(a^{-1}) = f(a)\bar{e}f(a)^{-1} = f(a)f(a)^{-1} = \bar{e}$$

and this completes the proof.

We now consider the special case of homomorphism, namely isomorphism and show that:

5.3.3 *Theorem*

The relation of isomorphism in the set of all groups is an equivalence relation.

Proof: If G and $\bar{G}$ are isomorphic we write this as $G \approx \bar{G}$. We have to show that:

(i) $G \approx G$ (reflexivity)

(ii) $G \approx \bar{G}$ implies $\bar{G} \approx G$ (symmetry),

(iii) $G \approx \bar{G}, \bar{G} \approx \bar{\bar{G}}$ implies $G \approx \bar{\bar{G}}$ (transitivity).

For each $x \in G$ define $f: G - G\ by\ f(x) = x$. Then f is an isomorphism of G onto itself and hence $G \approx G$.

Let $G \approx \bar{G}$. Then there exists an isomorphism f from G onto $\bar{G}$. Since f is 1–1 and onto, the map $f^{-1}: \bar{G} \to G$ exists and is 1-1 and onto. Let $a, b \in \bar{G}$ and let $f(x) = a, f(y) = b$ for $x, y \in G$. Now $x = f^{-1}(a), y = f^{-1}(b)$, $f(xy) = f(x)f(y) = ab$ and so $f^{-1}(ab) = xy = f^{-1}(a)f^{-1}(b)$. Then f^{-1} is also a homomorphism and so $\bar{G} \approx G$.

For the transitivity, let $G \approx \bar{G}$ and $\bar{G} \approx \bar{\bar{G}}$. Thus there exist isomorphisms f and g from G onto $\bar{G}$ and from $\bar{G}$ onto $\bar{\bar{G}}$respectively. Since f and g are one-one and onto, the composition g of : $G \to \bar{\bar{G}}$ is 1-1 and onto.

Further from $x, y \in G$.

$$(gof)(xy) = g(f(xy)) = g(f(x)f(y)) = g(f(x))g(f(y))$$

Thus gof is a homomorphism and hence $G \approx \bar{\bar{G}}$. This completes the proof.

Thus it follows from the above theorem that when two groups are isomorphic, then they are indistinguishable in some sense and we can treat them equal. They differ only in that their elements are labeled differently. If we know a given computation in one group, we can carry out the analogous computation in another group to which it is isomorphic.

5.3.4 *Theorem* *(Fundamental theorem of Homomorphism)*

Let φ be a homomorphism of G into $\bar{G}$, with kernel K. Then $G/K \approx \bar{G}$.

Proof

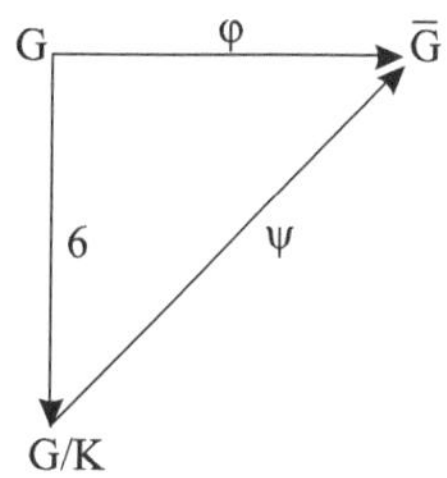

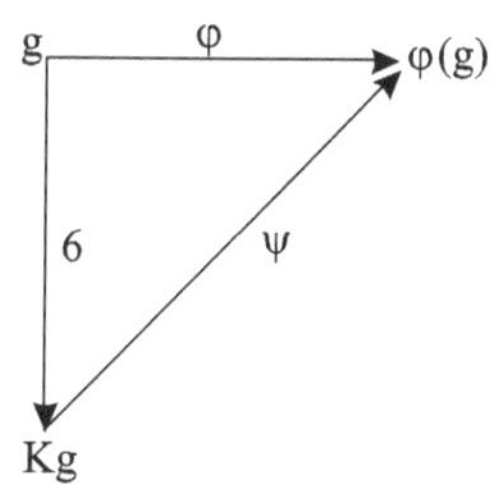

Define the mapping $\psi: G/K \to \bar{G}$ as follows, If $X \in G/K, X = Kg$ then $\psi(X) = \phi(g)$. First of all note that ψ is well defined,

If $X = Kg = Kg', g, g' \in G$

$\psi(X) = \phi(g)$ and $\psi(X) = \phi(g')$

In order that ψ makes sense it must be necessary that $\phi(g) = \phi(g')$. Suppose $Kg = Kg'$. Then

$$g = kg', k \in K.$$

Therefore $\phi(g) = \phi(kg') = \phi(k)\phi(g') = \bar{e}\phi(g') = \phi(g')$

is onto: For, if $\bar{x} \in \bar{G}, \bar{x} = \phi(g), g \in G$ since ϕ, is onto, we have

$$\bar{x} = \phi(g) = \psi(X) = \psi(Kg)$$

Thus ψ is onto. ψ is a homomorphism of $\frac{G}{K}$ onto $\bar{G}$. If $X, Y \in \frac{G}{K}, X = Kg$, $Y = Kf, g, f \in G$ then $XY = KgKf = Kgf$.

Hence

$\psi(XY) = \psi(K_{gf}) = \phi(gf) = \phi(g)\phi(f)$, since ϕ is a homomorphism of G onto $\bar{G}$. But

$$\psi(X) = \psi(Kg) = \phi(g) \text{ and } \psi(Y) = \psi(K_f) = \phi(f)$$

Therefore $\psi(XY) = \psi(X)\psi(Y)$

Finally we show that kernel of ψ = unit element in G/K, The unit element of $G/K = K_e = K$.

Let $Ker\psi = Kg$.

Now $\bar{e} = \psi(K_g) = \phi(g)$

So $g \in Ker\phi = K$. Hence $K_g = K$, and this completes the proof.

Remark

If N is a normal subgroup of G then it follows from above theorem that upto isomorphism G/N is a homomorphic image of G. Therefore there exists a 1–1 correspondence between homomorphic images of G and normal subgroups of G.

Let G be a group and let $A(G)$ be the set of all automorphisms of G onto itself.

Note that $A(G)$ is nonempty, for I the identity map, belongs to $A(G)$, we have.

5.3.5 *Theorem*

$A(G)$ is a group, the composition being the usual composition of maps.

Proof: First of all note that $A(G)$ is closed under the composition.

Let $T_1, T_2 \in A(G)$.

We shall show that

$$T_1, T_2 \in A(G).$$

Let $x, y \in G$. Since

$$T_1T_2 \in A(G), T_1(xy) = (T_1x)(T_1y) \text{and } T_2(xy) = (T_2x)(T_2y).$$

Now

$$T_1T_2(xy) = T_1(T_2(xy)) = T_1((T_2x)(T_2y)) = (T_1(T_2x))(T_1(T_2y))$$
$$= (T_1T_2(x))\,(T_1T_2(y))$$

We have only to show that if $T \in A(G)$ then $T^{-1} \in A(G)$.

Now if,

$$T((T^{-1}x)T^{-1}y) = T(T^{-1}x)(T(T^{-1}\,y)) = ((TT^{-1})(x))\,(TT^{-1}(y))$$
$$= I(x)(Iy) = xy$$

This means that,

$$T^{-1}(xy) = (T^{-1}x)(T^{-1}y)$$

and completes the proof.

5.4 DISCUSSIONS

We have just demonstrated that $A(G)$ the set of all automorphisims of a group G into itself is nonempty and is a group. It is now a pertinent question whether $A(G)$ is a nontrivial group or whether it consists of I only; in other words do there exist nontrivial automorphisims. If G is a group consisting of two elements only, then it is left to the reader as an exercise to verify that $A(G) = \{I\}$. But we shall now demonstrate that if G has more than two elements, then $A(G)$ is nontrivial.

Firstly, let G be abelian and let there exist an element $x_0 \in G$ such that,

$$x_0 \neq x_0^{-1}.$$

Let $T: G \rightarrow G$ be defined by $T(x) = x^{-1}$ for all $x \in G$. Then T is an automorphisim and $T \neq I$.

Now let G be any group, not necessarily abelian. For given $g \in G$ define a map by $Tg: G \rightarrow G$ by

$$T_g(x) = g^{-1}xg \text{ for all } x \in G$$

observe that T_g is an automorphism of G onto G. For given

$$y \in G, \text{ let } x = gyg^{-1}.$$

Then

$$\begin{aligned} T_g(x) &= g^{-1}xg = g^{-1}(gyg^{-1})g \\ &= (g^{-1}g)y(g^{-1}g) = y \end{aligned}$$

and hence T is onto.

If $T_g(x) = T_g(y)$ then $g^{-1}xg = g^{-1}yg$ and hence $x = y$. Thus Tg is 1–1.

Further for $x, y \in G$.

$$\begin{aligned} T_g(xy) &= g^{-1}(xy)g = g^{-1}(xgg^{-1}y)g \\ &= (g^{-1}xg)(g^{-1}yg) = T_g(x)T_g(y) \end{aligned}$$

Thus T_g is a homomorphism and hence it is an automorphism of G onto itself. If G is nonabelian, then there exists a pair of elements $a, b \in G$ such that $ab \neq ba$, then $T(a)b = a^{-1}ba \neq b$ and so $Ta \neq I$. We now make the following definition.

5.4.1 *Definition*

For an element g in a group, G the automorphism Tg from G onto G defined by $T_g(x) = g^{-1}xg$ is called the inner automorphism corresponding to g. We write $I(G)$ for the set of all inner automorphisms of G, i.e.

$$I(G) = \{T_g \in A(G): g \in G\}$$

We have.

5.4.1 *Theorem*

$I(G)$ is a group and $I(G) \approx G/Z$ where Z is the centre of G.

Proof: It is left to the reader to verify that $I(G)$ is a subgroup of $A(G)$ and hence $I(G)$ is a group.

To show that $I(G) = G/Z$ define a map $\psi: G \to A(G)$ by

$$\psi(g) = T_g$$

for every $g \in G$

Then it can be easily checked that ψ is a homomorphism from G into $A(G)$, whose image is $I(G)$. Now we claim that the Kernel of ψ is Z. Suppose it is K and let $g_0 \in K$. Then $\psi(g_0) = I$ i.e. $Tg_0 = I$. This means that for any $x \in G, Tg_0(x) = x$. However, $Tg_0(x) = g_0{}^{-1}xg_0$ and so $x = g_0{}^{-1}xg_0$ for all $x \in G$, i.e. $g_0x = xg_0$ for all $x \in G$. Thus, $K \subset Z$. Now if $z \in Z$, then $T_zx = z^{-1}xz = z^{-1}(zx) = x$, so $T_z = I$ and hence $z \in K$, showing that $Z \subset K$. Therefore, $K = Z$.

Observe that ψ is a homomorphism of G into $A(G)$ with image $I(G)$ and Kernel Z.

Now applying the Fundamental theorem of homomorphism we get the required conclusion.

5.4.2 *Theorem (Cayley)*

Every group is isomorphic to a subgroup of $A(S)$ for some appropriate S where $A(S)$ denotes the group of all 1–1 and onto mappings from a set S onto itself.

Proof: Let G be a group. Take $S = G$. For $g \in G$ define $T_g : S = G \to S = G$ by,

$$T_g(x) = xg$$

for every $x \in G$

If $y \in G$ then $y = (yg^{-1})g = T_g(yg^{-1})$, so that T_g is onto. If $x, y \in S$ and $T_g(x) = T_g(y)$, then $xg = yg$ and hence $x = y$. Therefore T_g, is 1–1, and hence for every $g \in G, T_g \in A(S)$. Now define a map $\psi: G \to A(S)$ by

$$\psi(g) = T_g$$

for all $g \in G$

Observe that ψ is a homomorphism. Let K be the kernel of ψ and let $g_0 \in K$.

Then

$\psi(g_0) = Tg_0$ is the identity map on S.

So for all $x \in G$, and in particular, for $e \in G$,

$$e = eTg_0 = eg_0 = g_0$$

Thus $g_0 = e$ and hence $K = \{e\}$.

Now it follows that ψ is an isomorphism of G into $A(S)$ and this proves the theorem.

EXERCISES

1. Prove that a group with the property that $x^2 = e$ for every element x is abelian.
2. Show that a group of order n with $n \leq 4$ is abelian.
3. Show that the group of all 1–1 maps of a nonempty set X onto itself is non-abelian.
4. Let G be a cyclic group of order n and H a cyclic group of order m. If m and n are relatively prime, then show that GXH is cyclic.
5. If G is an abelian group then for all $a, b \in G$ and all integers n show that $(ab)^n = a^n b^n$.
6. If G is a group such that $(ab)^2 = a^2b^2$ for all $a, b \in G$ then prove that G must be abelian.
7. If G is a finite group then show that there exists a positive integer N such that $a^N = e$ for all $a \in G$.
8. Show that if every element of a group G is its own inverse, then G is abelian.
9. If G is a group of even order then show that it has an element $a \neq e$ such that $a^2 = e$.
10. Let G be a finite set which is closed under an associative product and let both cancellation laws hold in G. Show that G is a group.
11. Let G be a group. If p is a prime number and $p \mid o(G)$ then show that G has an element of order p.
12. Let G be a finite abelian group and let the integer n be relatively prime to $o(G)$. Prove that every $g \in G$ can be written as $g = x^n$ with some $x \in G$.

    [*Hint:* Consider the map $\varphi: G \to G$ defined by $\varphi(y) = y^n$ and show that this mapping is an isomorphism of G onto G].

13. Show that if G is a cyclic group of order n and if d is a positive integer such that $d|n$ then G contains exactly one subgroup of order d and this subgroup is cyclic.
14. Let G be an abelian group and H the subset of elements of G of finite order then show that H is a normal subgroup of G.
15. If $G = S_3, H$ and K are Subgroups of G such that $o(H) = 3$ and $o(K) = 2$, then must HK be a subgroup of G?
16. Show that every subgroup of an abelian group is normal.
17. If G is a group and H a subgroup of index two in G then prove that H is a normal subgroup of G.
18. If a cyclic subgroup H of group G is normal in G, then show that every subgroup of H is normal in G.
19. If G is an abelian group and if N is any subgroup of G, then prove that G/N is abelian.
20. If H is the only subgroup of $o(H)$ in a finite group G, then show that H is a normal subgroup of G.
21. Show that the group of automorphisms of a cyclic group of order 8 is a non-cyclic group of order 4.
22. Verify whether the following mappings are homomorphisms:
    (a) Let G be the group of nonzero real numbers under multiplication and $\varphi: G \to G$ defined by
    $\varphi(x) = x^2$ for all $x \in G$
    (b) Let G be as in (i) and $\varphi: G \to G$ defined by
    $\varphi(x) = 2^x$ for all $x \in G$
    (c) Let G be the group of real numbers under addition and $\varphi: G \to G$ defined by
    $\varphi(x) = x + 1$ for all $x \in G$
    (d) Let G be as in (c) and $\varphi: G \to G$ defined by
    $\varphi(x) = 7x$ for all $x \in G$
    (e) Let G be any abelian group and $\varphi: G \to G$ defined by
    $\varphi(x) = x^5$ for all $x \in G$
23. Verify whether the following mappings are auto morphisms:
    (a) Let G be the group of integers under addition and $\varphi: G \to G$ defined by
    $\varphi(x) = -x$ for all $x \in G$
    (b) Let G be the group of positive real numbers under multiplication and $\varphi: G \to G$ defined by
    $\varphi(x) = x^2$ for all $x \in G$
    (c) Let G be a cyclic group of order 12 and $\varphi: G \to G$ defined by
    $\varphi(x) = x^2$

CHAPTER 6

Rings and Fields

6.1 DEFINITIONS AND EXAMPLES

6.1.1 *Definition*

Let R be a non empty set. Let there be two binary compositions, called addition and multiplication, defined on it. Then R is called a ring if it is:

1. an abelian group with respect to addition
2. a semigroup with respect to multiplication
3. the two distributive laws hold: for $a, b, c \in R$

   $a(b + c) = ab + ac$

   $(b + c)a = ba + ca$

If we write explicitly, then the definition of ring will read as follows: A nonempty set R with two binary operations, addition and multiplication, defined on it is called a ring if for all $a, b, c \in R$:

1. $a + b \in R$
2. $(a + b) + c = a + (b + c)$
3. There is an element $0 \in R$ (called zero element) such that $a + 0 = 0 + a = a$
4. There is an element $-a \in R$ such that $a + (-a) = (-a) + a = 0$
5. $a + b = b + a$
6. $ab \in R$
7. $a(bc) = (ab)c$
8. $a(b + c) = ab + ac$; $(b + c)a = ba + ca$

A ring R is said to be a ring with unit element (or a ring with identity) if there is an element $1 \in R$ called the multiplicative identity, such that:

$$a1 = 1a = a\,; \forall a \in R.$$

A ring R is called a commutative ring if the multiplication in R satisfies the commutative property, i.e. if:

$$ab = ba;\ \forall a, b \in R.$$

A ring R is called a division ring (or a skew field) if all its nonzero elements form a group under multiplication.

A commutative division ring is called a field. In other words a ring R is called a field if all its nonzero elements form an abelian group under multiplication.

If R is a commutative ring, then $0 \neq a \in R$ is said to be a zero divisor if there exists an element $0 \neq b \in R$ such that $ab = 0$.

A commutative ring is called an integral domain if it has no zero divisors.

6.2 EXAMPLES OF RINGS AND FIELDS

6.2.1 *Example*

(a) The following are rings with respect to usual addition and multiplication:

1. The singleton set consisting only of the number 0.
2. The set $\mathbb{Z}$ of all integers.
3. The set of all even integers.
4. The set $\mathbb{Q}$ of all rational numbers.
5. The set $\mathbb{R}$ of all real numbers.
6. The set $\mathbb{C}$ of all complex numbers.

It should be noted that the set in (2) is an integral domain and the sets in (4), (5) and (6) are in fact fields.

(b) Any ring of subsets of a set U is a ring with respect to addition and multiplication defined by $A + B = A \triangle B$ and $AB = A \cap B$. The fact that a ring of sets is actually a ring is the reason for the name ring of sets.

(c) Let m be a positive integer and $\mathbb{Z}_m = \{0, 1, \dots, m-1\}$. In $\mathbb{Z}_m$ define a sum $a + b$ and the product ab to be the remainders obtained when their usual sum and product are divided by m. Then $\mathbb{Z}_m$ is a ring with respect to the addition and multiplication defined as above. If m is any positive integer, then $\mathbb{Z}_m$ is not always an integral domain. For example, if $m = 6$, then observe that $2 \cdot 3 = 0$ and also $3 \cdot 4 = 0$.

(d) The set $\{a + b\sqrt{2} : a, b \in \mathbb{Z}\}$ is a ring under usual addition and multiplication of real numbers.

(e) Let $\mathbb{Z}(i) = \{a + ib : a, b \in \mathbb{Z}\}$. Note that a complex number of the form $a + ib$ where $a, b \in \mathbb{Z}$, is called a Gaussian integer and $\mathbb{Z}(i)$ is called the set of all Gaussian integers. $\mathbb{Z}(i)$ is a ring under the usual addition and multiplication of complex numbers. More precisely it is an integral domain called the domain if Gaussian integers.

(f) The set $\{a + b\sqrt{2} : a, b \in \mathbb{Q}\}$ is a ring under usual addition and multiplication of real numbers.

(g) Let $\mathbb{R}$ denote the field of all real numbers. Then $\mathbb{R} \times \mathbb{R}$ is a field with respect to addition and multiplication of real numbers defined as follows.

$(a, b) + (c, d) = (a + c, b + d);$

$(a, b) \cdot (c, d) = (ac - bd, ad + bc).$

Note that (0,0) and (1,0) are the additive and multiplicative identity respectively.

If $(a, b) \neq (0,0) \in \mathbb{R} \times \mathbb{R}$, then $(a, b)^{-1} = \left(\frac{a}{a^2+b^2}, \frac{-b}{a^2+b^2}\right)$ since $(a, b) \cdot \left(\frac{a}{a^2+b^2}, \frac{-b}{a^2+b^2}\right) = (1, 0).$

(h) The set $\mathbb{R} \times \mathbb{R}$ is a ring under addition and multiplication defined as follows:

$(a, b) + (c, d) = (a + c, b + d);$

$(a, b) \cdot (c, d) = (ac, bd)$

It is neither an integral domain nor a field. It is not an integral domain because,

$(1, 0) \cdot (0, 1) = (0, 0)$, but $(1, 0) \neq (0, 0)$ and $(0, 1) \neq (0, 0)$.

(i) The set of all n −rowed square matrices form a ring under addition and multiplication of matrices. This is not an integral domain. First observe that matrix multiplication is not commutative. For if $n = 2$ then,

$\begin{pmatrix} 1 & 0 \\ 0 & 0 \end{pmatrix}\begin{pmatrix} 0 & 1 \\ 1 & 0 \end{pmatrix} = \begin{pmatrix} 0 & 1 \\ 0 & 0 \end{pmatrix}$ but $\begin{pmatrix} 0 & 1 \\ 1 & 0 \end{pmatrix}\begin{pmatrix} 1 & 0 \\ 0 & 0 \end{pmatrix} = \begin{pmatrix} 0 & 0 \\ 1 & 0 \end{pmatrix}$.

Also, by taking $n = 2$ we observe that,

$\begin{pmatrix} 1 & 1 \\ 2 & 2 \end{pmatrix}\begin{pmatrix} -1 & 1 \\ 1 & -1 \end{pmatrix} = \begin{pmatrix} 0 & 0 \\ 0 & 0 \end{pmatrix}$.

This set is also not a skew field. Because any non-zero square matrix A need not have a multiplicative inverse. This holds if and only if A is non-singular.

(j) Let M be the set of all 2×2 matrices of the form $\begin{pmatrix} \alpha & \beta \\ -\bar{\beta} & \bar{\alpha} \end{pmatrix}$ where α, β are complex numbers and $\bar{\alpha}, \bar{\beta}$ are respectively the complex conjugates of α and β. Then M is skew-field under addition and multiplication of matrices. If $A = \begin{pmatrix} \alpha & \beta \\ -\bar{\beta} & \bar{\alpha} \end{pmatrix}$ is a nonzero element of M and if $D = \alpha\bar{\alpha} + \beta\bar{\beta}$, then the matrix $B = \frac{1}{D}\begin{pmatrix} \alpha & \beta \\ -\bar{\beta} & \bar{\alpha} \end{pmatrix}$ is the inverse of A and belongs to M.

We now introduce a special class of rings which are important and useful in various applications.

6.2.2 *Definition*

A ring R is said to be Boolean ring if $a^2 = a$ for every $a \in R$.

6.2.2 *Example*

(a) The set $\{0, 1\}$ under usual addition and multiplication.

(b) Any ring of sets under addition and multiplication defined by the symmetric difference and intersection property.

Worked Out Examples

6.2.3 *Example*

(a) If R is a Boolean ring, then

(i) $a + a = 0$; $\forall a \in R$

(ii) $a + b = 0$ implies $a = b$

For every $a \in R, a + a \in R$. Since R is a Boolean ring we have,

$$(a + a) + 0 = (a + a) = (a + a)^2 = (a + a)(a + a)$$
$$= a(a + a) + a(a + a)$$
$$= (a^2 + a^2) + (a^2 + a^2)$$
$$= (a + a) + (a + a)$$

By cancellation law we have $a + a = 0$.

Let $a, b \in R$ be such that $a + b = 0$. But since R is a Boolean ring $a + a = 0$.

Therefore $a + a = 0 = a + b$ and by cancellation law it follows that $a = b$.

(b) Prove that every Boolean ring R is commutative. Let R be a Boolean ring and let $a, b \in R$.

We have

$$a + b = (a + b)^2 = (a + b)(a + b)$$
$$= a(a + b) + b(a + b)$$
$$= a^2 + ab + ba + b^2$$
$$= (a^2 + b^2) + ab + ba$$
$$= (a + b) + (ab + ba).$$

Now by cancellation law it follows that,

$ab + ba = 0$

and hence by the previous example we have,

$ab = ba.$

Thus $R$is commutative.

6.3 IDEALS AND HOMOMORPHISMS'

6.3.1 *Definition*

Let R be a ring and let S be a nonempty subset of R. S is said to be a subring of R if under the composition in R, S itself forms a ring. It is clear that a subring S of a ring R is a subgroup of the additive group R and that it is a semigroup under multiplication. Therefore we can give an alternative definition for a subring and this can be stated as a theorem as follows.

6.3.1 *Theorem*

Let R be a ring and let $\Phi \neq S \subset R. S$ is a subring of R if and only if,

1. $a, b \in S$ implies $a - b \in S$
2. $a, b \in S$ implies $ab \in S$.

The proof of this theorem is easy and is left to the reader as an exercise.

6.3.1 *Example*

(a) Let R be a ring. Then R itself and the singleton set $\{0\}$, where 0 is the additive identity in R are subrings of R. These two are called trivial or improper subrings of R and other subrings, if exist, are called nontrivial or proper.

(b) The ring of all even integers is a subring of the ring $\mathbb{Z}$ of all integers under usual addition and multiplication. Note that $\mathbb{Z}$ is a ring with identity, but the ring of all even integers does not have the identity.

(c) In the ring $\mathbb{Z}_6$ (the ring of all integers mod 6) the set $S = \{0, 2, 4\}$ forms a subring.

(d) Let R be a ring. We define Cent R the centre of R as,

$\text{Cent}\, R = \{c \in R: cx = xc \text{ for all } x \in R\}.$

Then Cent R is a subring of R. To show this let $a, b \in \text{Cent}\, R$ and let $x \in R$ be arbitrary.

Then $ax = xa, bx = xb$. Now

$(a - b)x = ax - bx = xa - xb = x(a - b)$

and hence $a - b \in \text{Cent } R$. Also

$(ab)x = a(bx) = a(xb) = (ax)b = (xa)b = x(ab)$

thus $ab \in Cent\ R$ and this completes the proof.

(e) The intersection of two subrings of a ring R is also a subring of R. Let R_1 and R_2 be any two subrings of R. Since R_1 and R_2 are nonempty subsets of R each containing 0, $R_1 \cap R_2$ is nonempty subset of R. Let $a, b \in R_1 \cap R_2$. Then $a, b \in R_1$ and $a, b \in R_2$. Since R_1 and R_2 are subrings, $a - b, ab$ belong to R_1 and also to R_2. Therefore $a - b \in R_1 \cap R_2$ and also $ab \in R_1 \cap R_2$. This shows that $R_1 \cap R_2$ is a subring of R.

Based on our experience in the study of group theory it is now natural to ask whether we can carry over some analog to rings of the concept of normal subgroup. Once this is achieved, one would hope that this analog would lead to the concept of quotient ring in the same manner as quotient group of a group was constructed with the help of a normal subgroup. We now introduce the concept of an ideal for rings. This is the analog to rings of the concept of normal subgroups and the most natural definition is the following.

6.3.2 *Definition*

A nonempty subset I of a ring R is said to be an ideal of R if:

1. I is a subgroup of R under addition.
2. For every $i \in I$ and $r \in R$,
   $ir \in I$ and $ri \in I$.

This is infact called two sided ideal. If R is a commutative ring then it is enough to say that for $i \in I$ and $r \in R$, $ri \in I$. In any ring R we define a left or right ideal according as $ri \in I$ or $ir \in I$ for all $i \in I$ and $r \in R$. By an ideal we always mean two-sided ideal.

Every ideal is clearly a subring, but converse may not be true.

6.3.2 *Example*

(a) The ring $\mathbb{Z}$ of all integers is a subring of the ring (field), $\mathbb{Q}$ of all rational numbers, but $\mathbb{Z}$ is not an ideal in $\mathbb{Q}$.

(b) $\mathbb{Q}$ is a subring (sub field) of the field $\mathbb{R}$ of all real numbers, but $\mathbb{Q}$ is not an ideal in $\mathbb{R}$.

(c) The ring of all even integers is an ideal in the ring $\mathbb{Z}$ of all integers.

(d) Let M be the set of all 2×2 matrices with real entries. Then this is a ring with matrix addition and multiplication. The set of all matrices of the form $\begin{pmatrix} a & b \\ 0 & 0 \end{pmatrix}$ is a right ideal in M where as the set of all matrices of the form $\begin{pmatrix} a & 0 \\ b & 0 \end{pmatrix}$ is a left ideal in M.

(e) If I_1 and I_2 are any two ideals of a ring R then $I_1 \cap I_2$is an ideal in R.

(f) If I_1 and I_2 are any two ideals of a ring R then $I_1 \cup I_2$ may not be an ideal always in R.

6.3.2 *Theorem*

Let R be a ring with (multiplicative) identity and I a proper ideal in R. Then no element of I has a multiplicative inverse in R.

Proof: Suppose, to the contrary, that $a \in I$ has a inverse $a^{-1} \in R$. Since I is an ideal we now have,

$$aa^{-1} = 1 \in I.$$

Let $r \in R$. Then $1r = r \in I$ and hence $R \subset I$. Consequently $I = R$ and this contradicts the given hypothesis that I is a proper ideal of R. Therefore the result follows.

We shall now discuss the quotient of a ring with the ideal just we did in case of a group with the help of normal subgroups.

Let R be a ring and I an ideal in R. Let R/I denote the set of all distinct cosets of I in R which is obtained by considering I as a subgroup of Runder addition. In otherwords, $R/I = \{a + I : a \in R\}$ where $a + I = \{a + i : i \in I\}$. We now define addition and multiplication in R/I in a natural way as follows:

$$(a + I) + (b + I) = (a + b) + I;\ (a + I)(b + I) = (ab) + I.$$

6.3.3 *Theorem*

If I is an ideal in a ring R, then R/I is a ring under addition and multiplication defined as above. If R is commutative, then so is R/I and if R has a unit element, R/I has also a unit element.

Proof: By the result of Chapter 5 it follows that the addition defined above is meaningful and that R/I is group under addition. We shall now show that the multiplication defined above is meaningful. We shall, infact, show that if

$a + I = a' + I$ and $b + I = b' + I$ then $(a + I)(b + I) = (a' + I)(b' + I)$, equivalently, we have to show that $ab + I = a'b' + I$. Note that $a + I = a' + I$, therefore $a = a' + i_1$ for some $i_1 \in I$. Similarly $b + I = b' + I$, therefore $b = b' + i_2$ for some $i_2 \in I$. Now,

$$ab = (a' + i_1)(b' + i_2) = a'b' + a'i_2 + i_1b' + i_1i_2.$$

Since I is an ideal, $a'i_2, i_1b'$and i_1i_2 are all in I. Consequently, $a'i_2 + i_1b' + i_1i_2 \in I$. Then $ab = a'b' + i_3$ with $i_3 \in I$. From this we conclude that $ab + I = a'b' + I$. We leave it as an exercise for the reader to check that R/I is a ring. Further note that if R is commutative, then so is R/I for $(a + I)(b + I) = ab + I = ba + I = (b + I)(a + I)$. Also note that if R has unit element 1, then R/I has also unit element $1 + I$.

6.3.4 *Theorem*

Let R be a commutative ring with unit element 1 whose only ideals are $\{0\}$and R. Then R is a field.

Proof: We have only to show that for any $0 \neq a \in R$ there exists an element $0 \neq b \in R$ such that $ab = 1$. For this let $0 \neq a \in R$. Define the set $Ra = \{xa: x \in R\}$. Then Ra is an ideal in R. To show that, let $u, v \in Ra$ then $u = r_1a, v = r_2a$ for some $r_1, r_2 \in R$. Thus $u - v = r_1a - r_2a = (r_1 - r_2)a \in Ra$ and hence Ra is a subgroup of R under addition. Further, if $r \in R$ then $ru = r(r_1a) = (rr_1)a \in Ra$. Since R is commutative $ur = ru \in Ra$. This shows that Ra is an ideal in R.

Now by the hypothesis of the theorem, $Ra = \{0\}$ or $Ra = R$. Since $0 \neq a = 1a \in Ra$, $Ra \neq \{0\}$. Thus $Ra = R$. This means that every element of R is a multiple of a by some element of R. In particular, $1 \in R$ and hence it can be realized as a multiple of a. This means that there exists an element $b \in R$ such that $ba = 1$. This completes the proof of the theorem.

We shall now introduce two special types of ideals in ring and characterize them with the help of their quotient rings.

6.3.3 *Definition*

Let R be a ring. An ideal P is called *prime ideal* in R if $a, b \in R, ab \in P$ implies either $a \in P$ or $b \in P$. An ideal $M \neq R$ in R is said to be *maximal ideal* of R if whenever I is an ideal in R such that $M \subset I \subset R$ then either $I = R$ or $I = M$.

6.3.5 *Theorem*

If R is a commutative ring with unit element and P is an ideal in R, then P is a prime ideal if and only if R/P is an integral domain.

Proof: Let P be a prime ideal in R. First of all note that since R is a commutative ring with unit element, R/P is a commutative ring with unit element. Now let $a, b, \in R$ be such that $ab \in P$. Since P is a prime ideal it follows that either $a \in P$ or $b \in P$. This means that either $a + P = P$ or $b + P = P$. This shows that R/P has no nonzero zero divisor and hence it is an integral domain.

Conversely, let R/P be an integral domain. Let $a, b \in R$ be such that $ab \in P$. That is $(a + P)(b + P) = ab + P = P$. Since R/P has no nonzero zero divisor, either $a + P = P$ or $b + P = P$. That is either $a \in P$ or $b \in P$. Therefore P is a prime ideal and this completes the proof.

6.3.6 *Theorem*

If R is a commutative ring with unit element and M is an ideal in R, then M is a maximal ideal if and only if R/M is a field.

Proof: Let M be a maximal ideal in R. Since R is a commutative ring with unit element, R/M is also a commutative ring with unit element. Note that there is a 1–1 correspondence between the set of all ideals of R/M and the set of all ideals of R which contain M. Therefore R/M has only $\{0\}$ and itself as ideals. Thus R/M is a field.

Conversely, let M be an ideal in R and R/M a field. Since R/M is a field its only ideals are $\{0\}$ and R/M itself. But by the correspondence mentioned in the above paragraph, the ideal M of R corresponds to the ideal $\{0\}$ of R/M and R itself corresponds to R/M. Thus there is no ideal in R between M and other than these two. Therefore M is a maximal ideal and this completes the proof.

We have observed in the last chapter that homomorphism of group is an important and useful concept and plays an important role in the study of group theory. This suggests that the appropriate analog for rings could also lead to important ideas. With this motivation we now introduce the concept of homomorphism for rings. Since rings are equipped with two binary compositions a natural definition of homomorphism for rings would be a mapping which preserves both properties.

6.3.4 *Definition*

Let R and $\bar{R}$ be any two rings. A mapping $f: R \to \bar{R}$ is said to be homomorphism if for all $a, b \in R$

(i) $f(a+b) = f(a) + f(b)$

(ii) $f(ab) = f(a)f(b)$

A homomorphism of R into $\bar{R}$ is said to be an isomorphism if it is a 1–1 mapping. Two rings R and $\bar{R}$ are said to be isomorphic if there is an isomorphism of R onto $\bar{R}$.

If f is a homomorphism of a ring R into a ring $\bar{R}$ then the kernel of f, denoted by K_f is the set

$$K_f = \{a \in R: f(a) = \bar{0}, \text{the zero of } \bar{R}\}$$

Remark

(a) It should be noted that in the definition of homomorphism addition and multiplication occurring on the left side of (i) and (ii) are those of R and those occurring on the right are those of $\bar{R}$. We used the same notation for convenience and this should not bring any confusion.

(b) If we ignore the multiplication in both the rings R and $\bar{R}$ then a (ring) homomorphism from R into $\bar{R}$ is at least a homomorphism from R into $\bar{R}$ when we consider them as abelian groups under their respective additions. Therefore as far as addition is concerned, all the properties about homomorphism of groups are carried over to rings. In particular, we have the following result whose proof is left to the reader as an exercise.

6.3.7 *Theorem*

If f is a homomorphism of a ring R into a ring $\bar{R}$ then

(a) $f(0) = \bar{0}$

(b) $f(-a) = -f(a)$ for each $a \in R$.

6.3.3 *Example*

(a) Let R and $\bar{R}$ be any two rings. Define $f: R \to \bar{R}$ by $f(a) = \bar{0}$ for all $a \in R$. Then f is a homomorphism with $K_f = R$.

(b) Let R be any ring. Define $f: R \to R$ by $f(a) = a$ for all $a \in R$. Then f is a homomorphism with $K_f = \{0\}$.

(c) Let R be any ring and S a subring of R. The inclusion map $i: S \to R$ defined by $i(a) = a$ for all $a \in S$ is a homomorphism.

(d) Let R be the ring of integers. Then$f: R \to R$ defined by $f(x) = 2x$ for all $x \in R$ is not a (ring) homomorphism, although it is group homomorphism, since $f(ab) = 2ab \neq 2a\, 2b = f(a)f(b)$.

(e) Let $\mathbb{Z}(\sqrt{2}) = \{a + b\sqrt{2} : a, b \in \mathbb{Z}\}$. Then $\mathbb{Z}(\sqrt{2})$is a ring under usual addition and multiplication of real numbers. Define $f: \mathbb{Z}(\sqrt{2}) \to \mathbb{Z}(\sqrt{2})$ by $f(a + b\sqrt{2}) = a - b\sqrt{2}$. Then f is a homomorphism of $\mathbb{Z}(\sqrt{2})$ into $\mathbb{Z}(\sqrt{2})$ and its kernel $K_f = \{0\}$.

(f) Let $\mathbb{Z}(i) = \{a + bi : a, b \in \mathbb{Z}\}$. Then $\mathbb{Z}(i)$ forms a ring under usual addition and multiplication of complex numbers. Define $f: \mathbb{Z}(i) \to \mathbb{Z}(i)$ by $f(a + bi) = a - bi$. Then f is a homomorphism.

(g) Let $\mathbb{C}$ be the ring(field) of complex numbers. Define $f: \mathbb{C} \to \mathbb{C}$ by $f(z) = \bar{z}$. Then f is a homomorphism.

(h) Let $\mathbb{Z}$ be the ring of integers and $\mathbb{Z}_n$, the ring of integers modulo n. Define $f: \mathbb{Z} \to \mathbb{Z}_n$ by $f(a) =$ remainder of a on division by n. Then f is a homomorphism from $\mathbb{Z}$ onto $\mathbb{Z}_n$ and K_f consists of all multiple of n.

(i) Let $\mathbb{C}$ be the field of complex numbers and M_2 the ring of 2×2 matrices of real entries. Then the map $f: \mathbb{C} \to M_2$ defined by $f(a + bi) = \begin{pmatrix} a & b \\ -b & a \end{pmatrix}$ is an isomorphism.

6.3.8 *Theorem*

Let f be a homomorphism from a ring R into a ring $\bar{R}$ with kernel K_f. Then K_f is an ideal in R.

Proof: By results of Chapter 5 it follows that K_f is a normal subgroup of R. Again for all $r \in R$ and $a \in K_f$ we have $f(ra) = f(r)f(a) = f(r)\bar{0} = \bar{0}$, which implies that $ra \in K_f$ and this completes the proof.

6.3.9 *Theorem*

Let f be a homomorphism from R into $\bar{R}$. Then f is an isomorphism if and only if $K_f = \{0\}$.

Proof: If f is an isomorphism, clearly $K_f = \{0\}$. Conversely let f be a homomorphism and $K_f = \{0\}$. Suppose $a, b \in R, a \neq b, f(a) = c, f(b) = d$.

It is now enough to show that $c \neq d$. Now since $a \neq b, a - b \neq 0$ and hence $a - b \notin K_f$ since $K_f = \{0\}$. In other words $f(a-b) \neq 0$. But f is a homomorphism, so $f(a-b) = f(a) - f(b) = c - d \neq 0$ and hence $c \neq d$. This completes the proof.

We have the following homomorphism theorem for rings and the proof is exactly same as that for groups.

6.3.10 *Theorem*

Let R and $\bar{R}$ be rings and f a homomorphism from R into $\bar{R}$ with kernel I. Then $\bar{R}$ is isomorphic to R/I. Moreover, there is a 1–1 correspondence between the set of all ideals of $\bar{R}$ and the set of ideals of R which contain I. This correspondence is achieved by associating an ideal $\bar{I}$ in $\bar{R}$ with ideal I in R defined by $I = \{x \in R: f(x) \in \bar{I}\}$. With I so defined, R/I is isomorphic to $\bar{R}/\bar{I}$.

6.4 INTEGRAL DOMAIN AND FIELD OF QUOTIENTS

6.4.1 *Theorem*

Every field is an integral domain, but the converse is not necessarily true.

Proof: Since a field is a commutative ring with identity, in order to prove that it is an integral domain it is enough to show that it does not have a nonzero zero divisor. Let F be a field and $a, b \in F$ be such that $ab = 0$. Let $a \neq 0$. Then $a^{-1} \in F$ and hence $ab = 0$ implies that $0 = a^{-1}0 = a^{-1}(ab) = (a^{-1}a)b = 1b = b$. Thus $b = 0$ and this completes the proof.

To show that the converse is not always true, consider the ring of all integers under addition and multiplication. This is an integral domain. But it is not a field.

6.4.2 *Theorem*

Every finite integral domain is a field.

Proof: Let $D = \{a_1, a_2, \ldots, a_n\}$ be a finite integral domain. In order to prove that D is a field we have to:

(i) produce an element $1 \in D$ such that $1a = a$ for every $a \in D$ and
(ii) for every element $0 \neq a \in D$ produce an element $b \in D$ such that $ab = 1$.

For this let $0 \neq a \in D$. Then the elements $aa_1, aa_2, \ldots, aa_n$ are all in D and they all are distinct. For if, $aa_i = aa_j, i \neq j$, then $a(a_i - a_j) = 0$. Since D is an integral domain and $a \neq 0$, we must have $a_i - a_j = 0$ and hence $a_i = a_j$ which contradicts that $i \neq j$. Thus $aa_1, aa_2, \ldots, aa_n$ are all distinct elements of D which has exactly n elements. Hence they are precisely all the elements of D, in other words every element $a \in D$ is of the form $a = aa_{i_0}$ for some $a_{i_0} \in D$. Since D is commutative, $a = aa_{i_0} = a_{i_0}a$. Now if $y \in D$ then we have $y = aa_i$ for some $a_i \in D$ and so $ya_{io} = (aa_i)a_{io} = (a_ia)a_{io} = a_i(aa_{i_0}) = a_ia = y$. Thus a_{i_0} is a unit element for D and we write it as 1. Now $1 \in D$, and therefore there exists a $b \in D$ such that $ab = 1$. This completes the proof.

6.4.3 *Theorem*

If p is a prime number, then $\mathbb{Z}_p$, the ring of integers mod p is a field.

Proof: We shall, infact, show that $\mathbb{Z}_p$is an integral domain. Since it is a finite set, it would then follow from the previous theorem, that $\mathbb{Z}_p$ is a field.

Let $i, j \in \mathbb{Z}_p$ and $ij = 0$. Then p must divide ij. Since p is prime, it must divide i or j. Then either $i \equiv 0 (\text{mod } p)$ or $j \equiv 0 (\text{mod } p)$, i.e. either $\bar{\imath} = \bar{0}$ or $\bar{\jmath} = \bar{0}$. Thus $\mathbb{Z}_p$ is an integral domain and this completes the proof.

Although, as observed earlier, any integral domain is not a field, but given any integral domain D, we can always construct a field F which will contain properly an isomorphic image of D. Such a field F is called the field of quotients or the quotient field of the integral domain D and the integral domain D is said to be imbedded in the field F. In particular, if we start the construction with the ring of integers, then its quotient field we get is the field $\mathbb{Q}$ of all rational numbers. We now describe the construction.

6.4.4 *Theorem*

Every integral domain can be imbedded in a field.

Proof: Let D be the given integral domain. Let M be the set of all ordered pairs (a, b) where $a, b \in D$ and $b \neq 0$. In M define a relation $\sim$ as follows $(a, b) \sim (c, d)$ if and only if $ad = bc$. This defines an equivalence relation on M.

1. If $(a, b) \in M$ then $(a, b) \sim (a, b)$ since $ab = ba$ and hence $\sim$ is reflexive.
2. If $(a, b), (c, d) \in M$ and $(a, b) \sim (c, d)$, then $ad = bc$ hence $cb = da$ and so $(c, d) \sim (a, b)$. This shows that $\sim$ is symmetric.

3. If (a,b), (c,d), $(e,f) \in M, (a,b) \sim (c,d)$ and $(c,d) \sim (e,f)$, then $ad = bc$ and $cf = de$ hence $af = be$. Thus $(a,b) \sim (e,f)$ and the relation $\sim$ is transitive.

Hence $\sim$ divides M into mutually disjoint equivalence classes. Let $[a,b]$ be the equivalence class of (a,b) in M and let F be the set of all such equivalence classes $[a,b]$ where $a,b \in D$ and $b \neq 0$. This F is the field we are seeking. [In particular a rational number is an equivalence class $[a,b]$ of (a,b) where $a,b \in \mathbb{Z}$ and $b \neq 0$. One can also write a/b for rational number $[a,b]$ if there is no confusion. One should note that as ordered pairs $(1,2)$ and $(2,4)$ or as fractions $\frac{1}{2}$ and $\frac{2}{4}$ are distinct, but as rational numbers both are identical because both of them are in the same equivalence class.] It remains to show that the set F defined as above is a field. For this purpose we define addition and multiplication in F. We first dispose of the addition,

$$[a,b] + [c,d] = [ad + bc, bd].$$

Since $b \neq 0$ and $d \neq 0$, it follows that $bd \neq 0$ and hence $[ad + bc, bd] \in F$. We now assert that this addition is well defined, i.e. if $[a,b] = [a',b']$ and $[c,d] = [c',d']$ then $[a,b] + [c,d] = [a',b'] + [c',d']$. Since $[a,b] = [a',b']$ and $[c,d] = [c',d']$ we have $ab' = ba'$ and $cd' = dc'$. Now we have to show $[ad + bc, bd] = [a'd' + b'c', b'd']$. Now

$$\begin{aligned} (ad + bc)b'd' &= adb'd' + bcb'd' \\ &= ab'dd' + bb'cd' \\ &= ba'dd' + bb'dc' \\ &= bd(a'd' + b'c'). \end{aligned}$$

And this is the desired equality. Clearly $[0,b]$ acts as a zero element for this addition, $[-a,b]$ as the additive inverse of $[a,b]$ and F is a commutative group under this addition.

We now turn to the multiplication of F. We define

$$[a,b][c,d] = [ac, bd].$$

Since $b \neq 0$ and $d \neq 0$, it follows that $bd \neq 0$ and hence $[ac, bd] \in F$. In a manner very much similar to the case of addition one can show that if $[a,b] = [a',b']$ and $[c,d] = [c',d']$ then $[a,b][c,d] = [a',b'][c',d']$. One can show that the nonzero elements of F form a commutative group under multiplication where $[d,d]$ acts as the unit element and $[d,c]$ acts as the inverse of $[c,d]$ (note that since $c \neq 0, [d,c] \in F$). Also it is a routine verification that the distributive law holds in F and having established all these, we conclude that F is field.

All that remains to show that there is an isomorphism D into F. First notice that for $x \neq 0, y \neq 0$ in $D, [ax, x] = [ay, y]$ since $(ax)y = x(ay)$. Let us denote $[ax, x]$ by $[a, 1]$. Define $f: D \to F$ by $f(a) = [a, 1]$ for every $a \in D$. Clearly f is 1-1 and since,

$$f(a+b) = [a+b, 1] = [a, 1] + [b, 1] = f(a) + f(b)$$
$$f(ab) = [ab, 1] = [a, 1][b, 1] = f(a)f(b)$$

f is a homomorphism. Thus f is an isomorphism from D into F and this completes the proof.

EXERCISES

1. Let R be any ring and $a, b \in R$. Find an expression for $(a+b)^2$ and what is the form of this when R is commutative.
2. Let R be any ring and $a, b \in R$. Find an expression for $(a+b)^n$ where n is a positive integer. What is the form of this when R is commutative?
3. If I is an ideal in a ring R with identity 1 and if $1 \in I$ then prove that $I \neq R$.
4. Let R be a ring with identity, not necessarily commutative, such that the only right ideals of R are (0) and R. Prove that R is division ring.
5. If R is a ring with identity, show that any divisor of zero in R is singular.
6. Let R be the ring of all real functions defined on $[0,1]$. If X is a subset of $[0,1]$. Show that the set $I(X)$ in R defined by

   $I(X) = \{f: f(x) = 0 \text{ for every } x \in X\}$

   is an ideal. Show that this ideal is maximal if and only if X is singleton.
7. What are the regular elements, singular elements and divisors of zero in the ring of all subsets of a non empty set under symmetric difference and intersection? Under what condition is this ring a field?
8. In the ring of integers why 2 is neither regular nor singular?
9. Prove that any homomorphism of a field is either an isomorphism or takes each element to zero.
10. Prove that the set

    $\{a + b\sqrt{p}: a, b \in \mathbb{Q}, p \text{ is a prime}\}$

    is a field under the usual addition and multiplication.
11. Prove that if a ring R has a left identity and a right identity then both are equal.
12. An element a in a ring R is said to be idempotent if $a^2 = a$. Prove that the only idempotent elements of an integral domain with unity are 0 and 1.

13. An element a in a ring R is said to be nilpotent if $a^n = 0$ for some positive integer n. prove that only nilpotent element of an integral domain is the additive identity 0.
14. Let U is an ideal of a ring R and

    $r(U) = \{x \in R: xu = 0 \text{ for } u \in U\}$.

    Prove that $r(U)$ is an ideal in R.
15. Let U and V are ideals of a ring R and

    $U + V = \{u + v: u \in U, v \in V\}$.

    Prove that $U + V$ is also an ideal of R.
16. Show that every homomorphic image of a commutative ring is a commutative ring.
17. If R is a ring with unit element 1 and f is homomorphism of R onto R'prove that $f(1)$ is the unit element of R'.
18. Let R be a commutative ring with unity. Then prove that every maximal ideal of R is a prime ideal.
19. If R is a finite commutative ring with unity, prove that every prime ideal of R is a maximal ideal.
20. Find the principal ideal generated by 5 in the ring of integers $\mathbb{Z}$.
21. Prove that the quotient field of a finite integral domain coincides with itself.

CHAPTER 7

Boolean Algebras

In this chapter we discuss the concepts of Lattices and Boolean Algebras. Before doing this, we first discuss the concept of order relations.

7.1 PARTIAL ORDER RELATION

Let X be any nonempty set. A partial order relation (por) in X is a relation $\leq \subset X \times X$ such that for all $x, y, z \in X$

We have

1. $x \leq x$ (reflexive)
2. $x \leq y$ and $y \leq x \Rightarrow x = y$ (antisymmetric);
3. $x \leq y$ and $y \leq z \Rightarrow x \leq z$ (transitive)

We sometimes write $x \leq y$ in the equivalent from $y \geq x$. A non empty set X with a partial order relation defined on it, is called a partially ordered set (poset). We now give three important and simple examples of partially ordered sets.

7.1.1 *Example*

(a) Let X be the set of all positive integers, and $x \leq y$ means that x divides y.

(b) Let X be the set of all real numbers, and $x \leq y$ means that x is less than or equal to y in usual sense.

(c) Let X be the family of all subsets of universal set U and let $A \leq B$ mean $A \subset B$.

Two elements x and y in a poset are called comparable if either $x \leq y$ or $y \leq x$. Note that there may be pairs of elements in a poset which are not comparable. In Example (a), for example, the integers 3 and 7 are not comparable because neither divides the other; and in Example (c) if U has more than one element, there are always two subsets of U neither of which is a subset of the other.

Some partial order relations possess a fourth property in addition to the three required by the definition.

Any two elements are comparable.

A por with this property is called a total (or linear) order relation and a set with a total (or linear) order relation defined on it is called a totally ordered set or a linearly ordered set or a chain.

7.2 LATTICES AND BOOLEAN ALGEBRAS

Let $(X, \leq)$ be a poset and let $A \subset X$. An element $x \in X$ is said to be an upper bound for A if $a \leq x$ for every $a \in A$. x is said to be the least upper bound $(l.u.b)$ or supremum (if it exists) for A if x is an upper bound for A and if y is any other upper bound for A, then $x \leq y$. Similarly we define a lower bound and the greatest lower bound $(g.l.b)$ or infimum for a subset of a poset.

A poset $(X, \leq)$ is said to be a lattice if every two element subset of X (i.e., every pair of elements of X) has both the supremum and infimum.

If we define the infimum of x and y to be $x \wedge y$ and the supremum of x and y to be $x \vee y$, then these elements are uniquely detemined by x and y. The operations $\wedge$ and $\vee$ have the following properties.

Idempotent laws: $x \vee x = x, x \wedge x = x$
Commutative laws: $x \vee y = y \vee x, x \wedge y = y \wedge x$
Associative laws: $x \vee (y \vee z) = (x \vee y) \vee z, (x \wedge y) \wedge z = (x \wedge y) \wedge z$
Absorption laws: $x \vee (x \wedge y) = x, x \wedge (x \vee y) = x$

Each of these laws consists of two identities. In each case we prove only one identity. Observe that $x \leq x \vee x$. By reflexivity, $x \leq x$ and so $x \vee x \leq x$. Thus $x \wedge x = x$.

Now $y \leq x \vee y$ and $x \leq x \vee y$. Hence $y \vee x \leq x \vee y$.

Similarly $x \leq y \vee x$ and $y \leq y \vee x$. Therefore, $x \vee y \leq y \vee x$.

Thus by antisymmetry $x \vee y = y \vee x$.

Let $a = x \vee (y \vee z)$ and $b = (x \vee y) \vee z$. Then $x \leq a, y \vee z \leq a$, i.e., $y \leq a$ and $z \leq a$. Now $x \leq a$ and $y \leq a$ imply $x \vee y \leq a$.

Again $x \vee y \leq a$ and $z \leq a$ imply $b = (x \vee y) \vee z \leq a$. Similarly, it can be shown that $a = x \vee (y \vee z) \leq b$; and hence by antisymmetry $a = b$.

We have $x \leq \vee (y \wedge z)$. Also $x \wedge y \leq x$ and hence $x \vee (x \wedge y) \leq x$. Thus by antisymmetry.

$$x = x \vee (x \wedge y).$$

We now show that the above properties actually characterise lattices. We prove the following:

Let X be a nonempty set over which two operations $\wedge$ and $\vee$ are defined and let these operations satisfy the above four properties. Then a partial order relation can be defined on X such that X becomes a lattice in which $x \vee y$ and $x \wedge y$ are the supremum and infimum of x and y.

First, observe that $x \wedge y = x$ and $x \vee y = y$ are equivalent, for if $x \wedge y = x$ and $x \vee y = (x \wedge y) \vee y = (y \wedge x) \vee y = y$ and similarly, $x \vee y = y$ implies $x \wedge y = x$. We now define a relation $' \leq '$ on X as follows:

$x \leq y$ means that either $x \wedge y = x$ or $x \vee y = y$. Since $x \wedge x = x$ we have $x \leq x$ for each $x \in X$, so that the relation, $' \leq '$ is reflexive.

If $x \leq y$ and $y \leq x$, so that $x \wedge y = x$ and $y \wedge x = x$, then $x = x \wedge y = y \wedge x = y$. So that relation $' \leq '$ is anti-symmetric. Now if $x \leq y, y \leq z$, so that $x \wedge y = x$ and $y \wedge z = y$, then

$x \wedge z = (x \wedge y) \wedge z = x \wedge (x \wedge y) = x \wedge (y \wedge z) = x \wedge y = x$, so $x \leq x$ and hence the relation $' \leq '$ is transitive.

We now show that $x \wedge y$ is the infimim of x and y. Since $(x \wedge y) \vee x = x$ and $(x \wedge y) \vee y = (y \wedge x) \vee y = y$, we have $x \wedge y \leq x$ and $x \wedge y \leq y$.

If $z \leq x$ and $z \leq y$, so that $z \wedge x = z$ and $z \wedge y = z$, then $z \wedge (x \wedge y) = (z \wedge x) \wedge y = z \wedge y = z$, so $z \leq x \wedge y$. Thus $x \wedge y$ is the glb or infimum of x and y. In a similar vein, it can be shown that $x \vee y$ is the lub or supremum of x and y.

A lattice is said to be distributive if the following distributive laws holds,

$$x \vee (y \wedge z) = (x \vee y) \wedge (x \vee z)$$

$$x \wedge (y \vee z) = (x \wedge y) \vee (x \wedge z)$$

A Lattice is said to be complemented if it contains distinct elements 0 and 1 such that $0 \leq x \leq 1$

for every x and if each element x has a complement x' with the property that,

$$x \vee x' = 1, x \wedge x' = 0$$

A Boolean algebra is a complemented distributive lattice.

It is possible for an element of complemented lattice to have more than one complement. But in a Boolean algebra each element has a unique complement. To prove this assume that x^* is also an element with the property that $x \wedge x^* = 0$ and $x \vee x^* = 1$.

Then

$$x = x \wedge 1 = x \wedge (x \vee x') = (x \wedge x) \vee (x \wedge x')$$
$$= 0 \vee (x \wedge x') = x \wedge x'$$

So $x^* \leq x'$, similarly it can be shown that,

$$x' \leq x^* \text{ and so } x^* = x'.$$

Also it is clear that the complement of x' (complement of complement of x) is x . i.e.,

$$x'' = x.$$

Furthermore, $0 \wedge 1 = 0$ and $0 \vee 1 = 1$, so we have

$$0' = 1 \text{ and } 1' = 0$$

In a Boolean algebra we have

$$x \leq y \text{ implies } y' \leq x'.$$

Now if $x \leq y$, then $x \wedge y' \leq y \wedge y' = 0$ and hence

$$y' = y' \wedge 1 = y' \wedge (x \vee x') = (y' \wedge x) \vee (y' \vee x')$$
$$= 0 \vee (y' \wedge x') = y' \wedge x' .$$

Therefore $y' \leq x'$ and this proves that result.

Finally we show that in a Boolean algebra the following identities hold:

$$(x \wedge y)' = x' \vee y' \text{ and } (x \vee y)' = x' \wedge y'.$$

We only prove the first one. Observe that if $x' \leq z$ and $y' \leq z$, so that $z' \leq x$ and $z' \leq y$.

Then $z' \leq x \wedge y$ or $(x \wedge y)' \leq z$. This shows that $(x \wedge y)' \leq$ any upper bound of x' and y', so $(x \wedge y)' \leq x' \vee y'$. Therefore it remains to show that $x' \vee y' \leq (x \wedge y)'$. Now since $x \wedge y \leq x$ and $x \wedge y \leq y$ we have $x' \leq (x \wedge y)'$ and $y' \leq (x \wedge y)'$ and hence the result holds.

We may recall that a Boolean algebra is a complemented distributive lattice. To be more explicit a Boolean algebra $B = (B, \wedge, \vee, /, 0, 1)$ is a set B with two binary operations ($\wedge$, $\vee$), two universal bounds (0, 1) and one unary operation such that for all $x, y, z \in B$.

1. $x \wedge x = x, x \vee x = x$ (Idempotence)
2. $x \wedge y = y \wedge x, x \vee y = y \vee x$ (Commutativity)
3. $x \vee (y \vee z) = (x \vee y) \vee z, (x \wedge y) \wedge z = (x \wedge y) \wedge z$ (Associativity)
4. $x \vee (x \wedge y) = x, x \wedge (x \vee y) = x$ (Absorption)

5. $x \wedge [x \vee (x \wedge z)] = (x \wedge y) \vee (x \wedge z)$,
   $x \vee [x \wedge (x \vee z)] = (x \vee y) \wedge (x \vee z)$
6. $x \vee (y \wedge z) = (x \vee y) \wedge (x \vee z), x \wedge (y \vee z) = (x \vee y) \wedge (x \vee z)$ (distributivity)
7. $x \wedge 0 = 0, x \vee 0 = x$
   $x \wedge 1 = 1, x \vee 1 = 1$ (laws of the bounds)
8. $x \wedge x' = 0, x \vee x' = 1$ (laws of the complementation)
9. $(x')' = x$ (Laws of involution)
10. $(x \wedge y)' = x' \vee y', (x \vee y)' = x' \wedge y'$ (De Morgan Laws)

7.3 DUALITY PRINCIPLE

In united states all passenger cars have the following traffic regulations:

Cars always travel on the right of the street. On a multilane divided highway, left lanes are passing lanes. Cars may make left turns at a red light.

In England all passenger cars have the following traffic regulations:

Cars always travel on the left of the street. On a multilane divided highway, right lanes are passing lanes. Cars may make left turns at a red light.

This is an example of duality principle. Let us take a mathematical example. Let $(A, \leq)$ be a poset. If we define $(a \frac{\leq}{R} b)$ iff $b \leq a$, then $\left(A, \frac{\leq}{R}\right)$ is another poset. Further if $(A, \leq)$ is a lattice, then $\left(A, \frac{\leq}{R}\right)$ is also a lattice. The meet "$\wedge$" and join "$\vee$" operations with respect to $(A, \leq)$ will respectively be the join and meet operations with respect to $\left(A, \frac{\leq}{R}\right)$; in other words, the glb and lub for $(A, \leq)$ are respectively the lub and glb for $(A, \leq)$. This is a case of duality principle. Each of the laws stated above for lattices and Boolean algebras has two identities, one is the dual of the other.

7.4 BOOLEAN EXPRESSION

An element $x \in B$ is called a Boolean variable. If a Boolean variable takes two distinct values 0 and 1 (which denotes in particular the states "off" and "on" respectively of a switch or "No" and "Yes" respectively of a logical expression then it is called a binary Boolean variable. In applications usually binary Boolean variables are used and for convenience we call them simple Boolean variables.

A Boolean expression is an expression containing a finite number for symbols each representing a variable element and the operations $\wedge$, $\vee$, $'$. For example, $((x' \wedge y) \vee z) \vee (x \wedge y')$ is Boolean expression.

If $x_1, x_2, \ldots, x_n$ are Boolean variables, then we write $(x_1, x_2, \ldots, x_n)$ to denote a Boolean expression.

Let $(B, \vee, \wedge, ')$ be a Boolean algebra. To be more precise, a Boolean expression over this Boolean algebra is defined as follows:

(i) any element of A is a Boolean expression,

(ii) any variable name is a Boolean expression.

(iii) if e_1 and e_2 are Boolean expressions, then $e_1', e_1 \vee e_2, e_1 \wedge e_2$ are Boolean expressions.

Observe that if a Boolean expression and the values of the variables are given, we can evaluate the expression by substituting the values of the variables in the expression. For example, consider the Boolean expression.

$$(x_1 \vee x_2) \wedge (x_1' \vee x_2') \wedge (x_2 \vee x_3)'$$

Or the Boolean algebra $(\{0, 1\}\wedge,\vee, ')$.

If $x_1 = 0, x_2 = 1, x_3 = 0$, then the above expression becomes

$$(0 \vee 1) \wedge (0' \vee 1') \wedge (1 \vee 0)' = 1 \wedge 1 \wedge 0 = 0$$

Observe that Boolean expressions can be simplified to a great extent. For example, consider the following expression.

$$(a \vee b \vee c \vee d') \wedge (a \vee b \vee c \vee d) \wedge (a \vee b \vee b') \wedge (a \vee b)$$

To simplify this let

$$a \vee b \vee c = x.$$

Then

$$(a \vee b \vee c \vee d') \wedge (a \vee b \vee c \vee d) = (x \vee d') \wedge (x \vee d)$$

$= x \wedge (x \vee d) \vee d' \wedge (x \vee d\,)$ (distributive)

$= x \vee d' \wedge x \vee d' \wedge d$ (absorption and distributive)

$= (x \vee d' \wedge x) \vee 0$ (associative and law of bounds)

$= x$ (absorption and law of bounds)

Thus the given expression is given by

$$(a \vee b \vee c) \wedge (a \vee b \vee c') \wedge (a \vee b')$$

Put $a \vee b = y$. Then

$$(a \vee b \vee c) \wedge (a \vee b \vee c') = (y \vee c) \wedge (y \vee c')$$
$$= y \wedge (y \vee c) \vee c \wedge (y \vee c') \text{ (distributive)}$$
$$= y \vee c \wedge y \vee c \wedge c' \text{ (absorption and distributive)}$$
$$= (y \vee c \wedge y) \vee 0 \text{ (associative and law of bounds)}$$
$$= y \text{ (absorption and law of bounds)}$$

Hence the given expression is,

$$(a \vee b) \wedge (a \vee b')$$
$$= a \wedge (a \vee b') \vee b \wedge (a \vee b') \text{ (distributive)}$$
$$= a \vee b \wedge a \vee b \wedge b' \text{ (absorption and distributive)}$$
$$= (a \vee b \wedge a) \vee 0 \text{ (associative and law of bounds)}$$
$$= a \text{ (absorption and law of bounds)}$$

The algebra of logic (the collection of statements) is a Boolean algebra with respect to disjunction, conjunction and negation and in this case the universal bounds are tautology and the contradiction.

We can apply the laws of Boolean algebra to algebra of logic so that we can simplify complicated statements into simple ones. We can even avoid drawing truth tables which are sometimes complicated. For example, take

Let $p = (\neg p \wedge \neg q) \vee (p \vee q) \wedge (\neg p \vee q) \wedge q)$

Since $\neg(p \vee q) = \neg p \wedge \neg q$ (by De Morgan laws)

We have

$$(\neg p \wedge \neg q) \vee (p \vee q) = \neg(p \vee q) \vee (p \vee q)$$
$$= \text{tautology (by laws of complementation)}$$

Hence $p = \text{tautology} \wedge ((\neg p \wedge p) \wedge q)$
$= (\neg p \wedge p) \wedge q$ (law of bounds)
$= q$ (absorption law)

The reader is advised to verify this by truth table.

We shall now discuss the application of Boolean algebra to switching circuits.

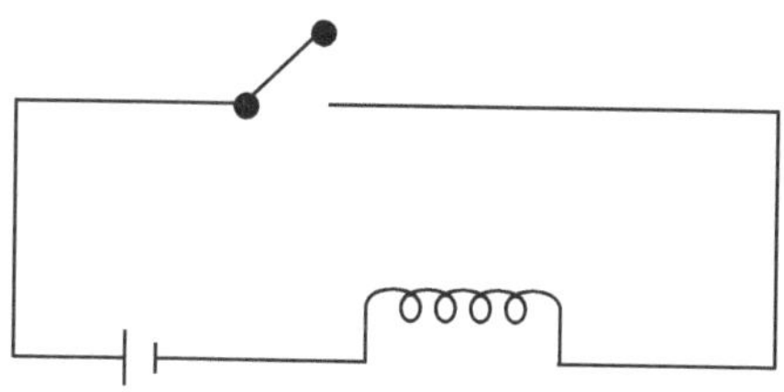

Fig. 1

A switching circuit consists of a battery, a bulb, some switches and connecting wires. It has two states: on or off. When the switch is on, the current flows and the bulb shines and when the switch is off, the current does not flow and the bulb does not shine. The value 1 or the symbol T is associated with the 'on' state of the switch and the value o or the letter F denotes the 'off' state of the switch. The switches in a circuit are denoted by the alphabets $a, b, c, \ldots, x, \ldots$ etc. Note that when two switches in a circuit are both 'on' or both 'off ', then we denote them by the same letter and if one switch is 'on' and the other is 'off', them we denote one by a and the other by a' (Complement of a).

7.5 SERIES AND PARALLEL SWITCHES

Two switches in circuit can be connected either in series or in parallel.

The following two figures show the series and parallel connections.

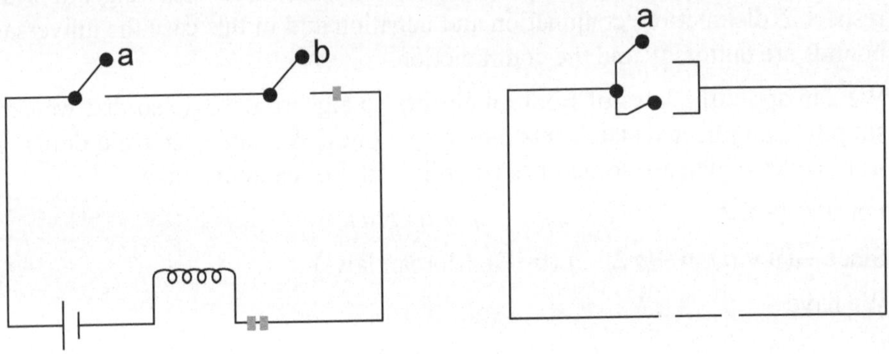

Fig. 2

The series and parallel connections of two switches a and b can be denoted by $a \wedge b$, and $a \vee b$ respectively and this can be represented in the tabular form,

Series Connections

a	a	$a \wedge b$
1	1	1
0	1	0
0	0	0
1	0	0

Parallel Connections

a	b	$a \vee b$
1	1	1
1	0	1
0	0	0
0	1	1

Simplification of switching circuits.

7.5.1 *Example*

Simplify the following circuit.

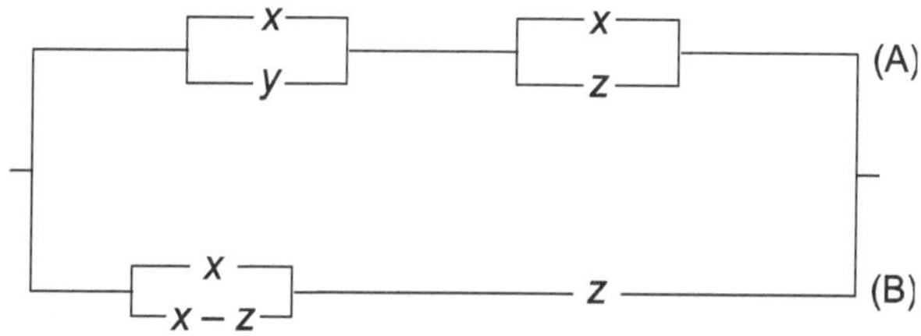

Fig. 3

We will first calculate the Boolean function which represents the above circuit. The circuit has two branches (A) and (B).

For (A):
The switches x and y are parallel, hence it is $x \vee y$. The switches x and z are parallel, hence it is $x \vee z$,

The above two are in series, hence the representation of (A) is,

$$(x \vee y) \wedge (x \vee z).$$

For (B):
x and z are in series, hence we have $x \wedge z$, the above is parallel to x, hence $x \vee (x \wedge z), x \vee (x \wedge z)$ is in series with x.

So we have

$$\big(x \vee (x \wedge z)\big) \vee z.$$

(A) and (B) are parallel. Hence the Boolean function for the whole circuit is,

$$(x \vee y) \wedge (x \vee z) \vee \big(x \vee (x \wedge z)\big) \wedge z$$

$$= x \wedge (x \vee z) \vee y \wedge \big(x \vee z) \vee (x \wedge z)\big)$$

$$= x \vee (x \wedge z) \wedge y \vee \big(x \vee z) \wedge (x \vee z)\big)$$

$$= x \vee (x \wedge z) \vee (y \vee z)$$

$$= x \vee (y \wedge z)$$

7.5.2 *Example*

Simplify the following network.

There are two parts in this circuit, the first part has representation,

$$x \vee y \ \vee (a \wedge b \wedge c)$$

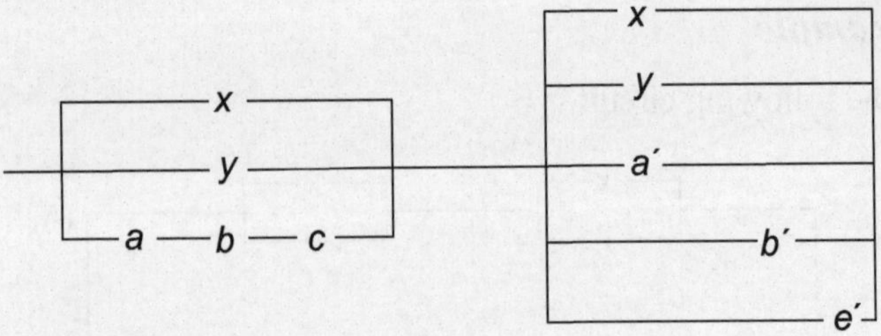

Fig. 4

and the second has the representation,

$$(x \vee y \vee (a \wedge b \wedge c)) \wedge (x \vee y \vee a' \vee b' \vee c')$$
$$= (x \vee y, \vee (a \wedge b \wedge c)) \wedge (x \vee y \vee (a \vee b \vee c)')$$
$$= (x \vee y) \wedge (x \vee y) \vee (a \wedge b \wedge c) \wedge (x \vee y) \wedge (x \vee y) \wedge (a \wedge b \wedge c)' \vee (a \wedge b \wedge c) \wedge (a \wedge b \wedge c)$$
$$= [(x \vee y) \vee (a \wedge b \wedge c) \vee (x \vee y)] \vee (x \vee y) \wedge (a \wedge b \wedge c)' \vee 0$$
$$= (x \vee y) \vee (x \vee y) \wedge (a \wedge b \wedge c)'$$
$$= (x \vee y)$$

The circuit represented by $x \vee y$ is

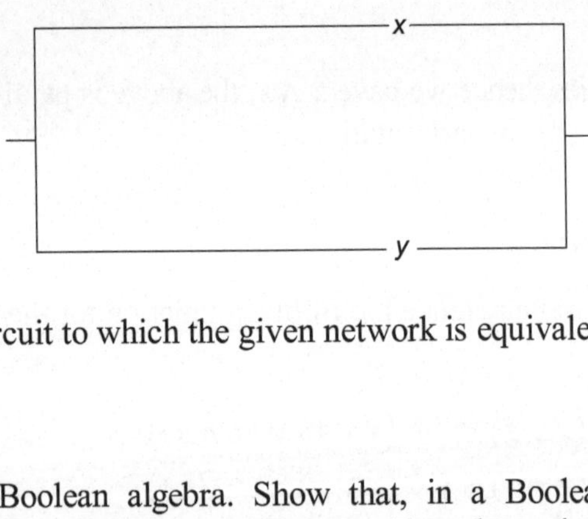

and this is the circuit to which the given network is equivalent.

EXERCISES

1. Define a Boolean algebra. Show that, in a Boolean algebra, every element has a unique complement. Also show that if x and y are elements of a Boolean algebra then the identities:

   $(x \wedge x)' = x' \vee y'$ and $(x \vee y)' = x' \wedge y'$ are true.

2. Let L be a lattice, if x, y, z are elements of L, verify the following:

   $x \wedge x = x, x \vee x = x, x \wedge y = y \wedge x, x \vee y = y \vee x$

   $x \wedge (y \wedge z) = (x \wedge y) \wedge z, x \vee (y \vee z) = (x \vee y) \vee z,$

   $(x \wedge y) \vee x = x, (x \vee y) \wedge x = x.$

3. Let L be a nonempty set in which two operations $\wedge$ and $\vee$ are defined which satisfy the following conditions:

$$\begin{aligned} &x \wedge x = x, && x \vee x = x; \\ &x \wedge y = y \wedge x, && x \vee y = y \vee x; \\ &x \wedge (y \wedge z) = (x \wedge y) \wedge z, && x \vee (y \vee z) = (x \vee y) \vee z; \\ &(x \wedge y) \wedge x = x, && (x \vee y) \wedge x = x. \end{aligned}$$

Show that L is a partially ordered set in which any two elements have a greatest lower bound and a least upper bound.

CHAPTER 8

Probability Theory

8.1 INTRODUCTION

In many physical phenomena, the result of each performance depends on "chance" and therefore cannot be predicted.

Even under apparently identical conditions, many observed quantities vary in an uncertain way. For example, the number of heads obtained in ten tosses of a coin varies unpredictably with each group of tosses, the total annual rainfall in Kharagpur varies unpredictably from year to year, and the total number of phone calls arriving at a switch board varies unpredictably from day to day. In this chapter we shall study probability theory. Probability theory is concerned with making rational statements about phenomena that are subject to uncertainty.

By "probability" we mean simply a number assigned to an event to measure the "likelihood" of that events happening. Intuitively we know the examples of events; the precise definition will be given shortly. Before going to invent the probability numbers which will be assigned to events we first observe what happens when we toss a coin.

If we toss a coin once, it is impossible to predict in advance whether the outcome will be a head or a tail. On the other hand if we toss the coin over and over again, something rather interesting occurs. To illustrate, consider Table 1. In this table we have recorded what happened at various stages when a coin was tossed 20,000 times. In early stages of tossing, the proportion of heads

Table 1

No. of Tosses (n)	*No. of Heads (h)*	*Proportion of Tosses h/n*
100	61	0.6100
1000	487	0.4870
5000	2550	0.5100
10,000	5097	0.5097
15,000	7649	0.5099
20,000	10,096	0.5048

varied rather considerably but as the coin tossing continued, the proportion began to settle down. These data suggest that if we continued coin tossing a large number of times, beyond 20,000, the proportion of heads would be very close to 0.5.

We now generalize the above idea to assign probability numbers to events. Suppose we have an experiment, the term experiment being used to mean the observation of any physical phenomena, and we want to observe the "likelihood" of an event's happening. Suppose the experiment can be repeated indefinitely under fixed conditions, and suppose during n repetitions of this experiment, a certain event occurs m times. We shall call the ratio m/n the relative frequency of the event after n repetitions. If this relative frequency approaches a number p as n gets larger and larger, then we shall call p, the probability of the event. This method of assigning probability numbers is not precise, since the phrase relative frequency approaches a number p has not been defined.

We can interpret the above fact in the following way, which is more precise.

If a coin is tossed once, the number 0.5, the probability of getting head, is in fact the ratio,

$$\frac{\text{the number of ways a head can occur}}{\text{the total number of outcome}}$$

If we generalize this fact, we can call the ratio,

$$\frac{\text{number of ways an event can occur}}{\text{the total number of possible outcomes}}$$

the probability of the event. This method of assigning probability numbers to events is called "The frequency ratio method".

There are also other methods of obtaining probability numbers such as: by intuition, with opinion polls, surveys etc. In some situations probability numbers can be obtained by intuition. For example, if we toss a coin, then intuition tells us that the probability of tossing a head is 1/2; if we have an ordinary six-sided die, then intuition tells us that the probability of tossing a 3 with this die is 1/6. We obtain these conclusions by arguing that each of the possible outcomes is equally likely and in this case intuition agrees with the frequency-ratio method of obtaining probability numbers.

On the other hand some probabilities cannot be obtained by using intuition; they can only be estimated from experimental data. For example, suppose we have a certain production process for manufacturing reels of magnetic tape for a digital computer and we are interested in the probability that the process will produce a defective reel. Intuition will not tell us what this probability is. If we test 20,000 reels and find that four of them are defective, then we can estimate the probability of a defective reel to be,

$$\frac{4}{20,000} = 0.0002.$$

Throughout this chapter we shall be concerned with outcomes of experiments, where the term experiment is used to mean the observation of any physical phenomena, and in probabilities of "events" associated with these experiments. Intuitively we know examples of events: the event that a head is tossed with a coin, the event that an even number is tossed with a six-sided die, the event that more than 100 cars enter a certain tunnel every day, the event that it will rain this evening and so on.

We shall now clearly identify all possible outcomes of an experiment and give a precise definition of an event.

The set of all possible outcomes of an experiment is called the sample space for the experiment; the elements of the sample space are called the sample points.

An event is a subset of the sample space of an experiment.

If S is the sample space of an experiment, then an event E is called a certain event if $E = S$; an impossible event if $E = \varphi$ and an elementary event if E consists of a single sample point. A certain event always occurs and an impossible event never occurs.

Since events are sets, given events E and F in a sample space S, we can form new events $E \cap F$, $E \cup F$ and E'.

The event $E \cap F$ consists of those sample points common to both E and F, in other words, $E \cap F$, consists of those sample points for which E and F both occur.

Two events E and F are said to be mutually exclusive if they cannot occur simultaneously i.e. if $E \cap F = \varphi$.

The notion mutually exclusive can be extended to more than two events. We call n events, $E_1, E_2, ..., E_n$ mutually exclusive if no two of them have any sample point in common i.e. if $E_i \cap E_j = \varphi$, $i \neq j$, Observe that n events are mutually exclusive means no two of them can occur simultaneously.

The event $E \cup F$ consists of those sample points that belong to E or to F or to both i.e., $E \cup F$ consists of those sample points for which E occurs or F occurs or both.

The event E' consists of those sample points that donot belong to E, i.e., E' consists of those sample points for which E does not occur.

Given a sample space, we are interested in assigning probabilities to various events. But, in practice, it is a tedious task, sometimes impossible, to assign probabilities to all the important events. In order to overcome this difficulty we assign probabilities only to certain basic events and then use various axioms and theorems to calculate the probabilities of other events as they are needed. Once a procedure is specified for obtaining the probabilities of all the events of interest, we say that a probability model has been assigned to the sample space.

We shall be concerned only with finite sample spaces and we shall assign probability numbers only to the elementary events. Then the probabilities of all other events can be obtained by using the following axioms:

1. $0 \le P(E) \le 1$ for any event $E \subseteq S$
2. $P(S) = 1$
3. $P(A \cup B) = P(A) + P(B)$ if $A \cap B = \varphi$.

Now the only problem before us is to assign a probability number to each elementary event of a finite sample space. Sometimes the probabilities of elementary events are determined by intuition and sometimes they can only be estimated by experimentation.

If the sample space of an experiment has k equally likely outcomes then each elementary event is assigned the probability $1/k$. This assignment of probabilities is called the uniform probability model for the experiment.

8.2 BASIC THEOREMS OF PROBABILITY

Theorem 1: If $A_1, \ldots, A_n$ are mutually exclusive events, then

$$P\left(\bigcup_{i=}^{n} A_i\right) = \sum_{i=1}^{n} P(A_i)$$

Proof: From the axioms the theorem is true for $n = 2$. Let the theorem be true for $n = m - 1$.

Then,

$$P\left(\bigcup_{i=1}^{m} A_i\right) = P\left(\bigcup_{i=1}^{m-1} A_i \cup A_m\right) = P\left(\bigcap_{i=1}^{m-1} A_i\right) + P(A_m)$$

$$= \sum_{i=1}^{m-1} P(A_i) + P(A_m) = \sum_{i=1}^{m} P(A_i).$$

Theorem 2: If E and F are events in a finite sample space,

Then,

$$P(E \cup F) = P(E) + P(F) - P(E \cap F).$$

Proof: The set $E \cup F$ can be expressed as,

$$E \cup F = (E \cap F') \cup (E \cap F) \cup (E' \cap F).$$

Since the three sets on the right hand side of the above equation are mutually exclusive, it follows from Theorem 1 that,

$$P(E \cup F) = P(E \cap F') + P(E \cap F) + P(E' \cap F).$$

But

$$E = (E \cap F') \cup (E \cap F)$$

and

$$F = (E' \cap F) \cup (E \cap F),$$

so that

$$P(E) = P(E \cap F') + P(E \cap F)$$

and

$$P(F) = P(E' \cap F) + P(E \cap F).$$

Therefore,

$$P(E \cup F) - P(E) - P(F) = -P(E \cap F)$$

and this completes the proof.

Corollary: If E and F are events in a sample space, then

$$P(E \cup F) \leq P(E) + P(F).$$

Theorem 3: If E is an event in a finite sample space S,

Then $P(E') = 1 - P(E)$.

Proof: Since $E \cup E' = S$ and $E \cap E' = \varphi$, we have

$$\begin{aligned} 1 &= P(S) \\ &= P(E \cup E') \\ &= P(E) + P(E'). \end{aligned}$$

Thus $P(E') = 1 - P(E)$.

Theorem 4: Probability is monotone, i.e. if $A \subset B$, then $P(A) \leq P(B)$ where A and B are events in a finite sample space.

Proof: If $A \subset B$, then $A^c = B^c \cup (B - A)$ and B^C and $B - A$ are disjoint.

Therefore,

$$P(B^c)+P(B-A)=P(A^c)$$

i.e., $1-P(B)+P(B-A)=1-P(A)$

i.e., $P(A)=P(B)-P(B-A)$

i.e., $P(A)\leq P(B)$.

Theorem 5: Probability is subtractive, i.e., if A and B are events in a finite sample space, then $P(B-A)=P(B)-P(A)$ where $A\subset B$.

Proof:

$$\begin{aligned}P(B-A) &= P(B\cap A^c)\\ &= P(B)+P(A^c)-P(B\cup A^c)\\ &= P(B)+1-P(A)-P(S)\\ &= P(B)+1-P(A)-1\\ &= P(B)-P(A).\end{aligned}$$

8.3 CONDITIONAL PROBABILITY

We know that the probability of tossing one with a die is 1/6. Now suppose that a die is tossed and we are given that the number on the top face is odd. What is the probability that the number tossed is a one? Since the number tossed must be odd, we have three equally likely outcomes for the experiment, that is, 1, 3 or 5.

Thus the probability that the number tossed is one, given that the number is odd is 1/3. This example shows that the probabilities can be affected by information about the outcome of the experiment. In this section we shall develop methods for computing probabilities when additional information about the outcome of an experiment is known.

In the tossing experiment discussed above, let E and F be respectively the events: an even number is tossed and an odd number is tossed. Then,

$$P(E) = 1/6.$$

But

$$P(E \text{ occurs given that } F \text{ occurs}) = 1/3 \qquad \dots (1)$$

In probability problems we write "E occurs given that F occurs" as E/F.

With this notation we can write (1) as,

$$P(E/F)=\frac{1}{3}.$$

The probability $P(E/F)$ is called the conditional probability of E given F.

Let S be a finite sample space and let A and B be two events associated with S. Our next objective is to formulate a relationship between $P(A/B)$ and the probabilities $P(A)$ and $P(B)$. In order to motivate this idea, consider the simplest case, an experiment with n equally likely outcomes. Suppose that A and B are two events in the sample space S. Let the event B contain l sample points and let the event $A \cap B$ contain m sample points. Then

$$P(B) = \frac{l}{n} \text{ and } P(A \cap B) = \frac{m}{n}.$$

We are now interested to calculate $P(A/B)$. Sine it is given that B occurs, the outcome of the experiment must be one of the l sample points in B. Of these l sample points, there are m points for which A occurs; these are the m sample points in $A \cap B$. Now since all outcomes are equally likely,

$$P(A/B) = \frac{m}{l}.$$

Thus

$$\begin{aligned} P(A/B) &= \frac{m}{l} = \frac{m/n}{l/n} \\ &= \frac{P(A \cap B)}{P(B)} \end{aligned} \qquad \text{... (2)}$$

Note that,

$$P(B/A) = \frac{P(B \cap A)}{P(A)} = \frac{P(A \cap B)}{P(A)}.$$

Therefore,

$$P(B/A) \neq P(A/B).$$

The formula (2) has been derived under the assumption that the outcomes are equally likely. But this situation motivates us for the following definition for the general case.

If A and B are events and $P(B) \neq 0$, then we define the conditional probability of A given by B by,

$$P(A/B) = \frac{P(A \cap B)}{P(B)}.$$

If A and B are two events sometimes it happens that the probability of A is unaffected by the additional knowledge that B occurs. To illustrate this, consider the following example.

8.3.1 *Example*

1. If a coin is tossed twice, find the probability that:
   (a) the second toss is a head,
   (b) the second toss is a head given that the first one is a head.

Let
A = the event that the second toss is a head
B = the event that the first toss is a head.

Thus, the sample space S has four points, A and B have two points each and the event $A \cap B$ has exactly one sample point that is, both the tosses are heads. Therefore, assuming that all the outcomes are equally likely we have,

$$P(A) = \frac{2}{4} = \frac{1}{2},$$

$$P(B) = \frac{1}{2} \text{ and } P(A \cap B) = \frac{1}{4}.$$

Now $P(A/B) = -\dfrac{P(A \cap B)}{P(B)} = \dfrac{1/4}{1/2} = \dfrac{1}{2}.$

The above example shows that there are events which do not influence the probabilities of one another. We shall now give a special name for such events.

Suppose that A and B are events such that,

$$P(A/B) = P(A)$$

or equivalently,

$$\frac{P(A \cap B)}{P(B)} = P(A).$$

This is equivalent to,

$$P(A \cap B) = P(A)P(B).$$

We now define the two events A and B are called independent if,

$$P(A \cap B) = P(A)P(B).$$

Note that if A and B are independent events with nonzero probabilities, then

$$P(A/B) = P(A) \text{ and } P(B/A) = P(B).$$

The notion of independence can be extended to more than two events as follows:

A set of events is called independent if each finite subset $\{A_1, A_2, \ldots, A_n\}$ satisfies,

$$P(A_1 \cap A_2 \cap \ldots A_n) = P(A_1)P(A_2)\ldots.P(A_n).$$

8.4 BAYES' THEOREM

If A and B are two events associated with a sample space, then in any practical problem it is necessary to know both the conditional probabilities $P(A/B)$ and $P(B/A)$. Bayes' Theorem gives a relationship between these two conditional probabilities and helps us to calculate one when the other is given. In 1763, Thomas Bayes (1702–1763) derived the following important result connecting $P(A/B)$ and $P(B/A)$.

$$P(B/A) = \frac{P(A/B)P(B)}{P(A)} \qquad \text{... (3)}$$

This is called Bayes' theorem.

Bayes' theorem has many varied and interesting applications.

With the use of set theory, the proof of Bayes' theorem is much easier for us than it was for Bayes.

Proof of Bayes' Theorem

By definition,

$$P(B/A) = \frac{P(B \cap A)}{P(A)}$$

$$= \frac{P(A \cap B)}{P(A)} \qquad \text{... (4)}$$

But $P(A/B) = \dfrac{P(A \cap B)}{P(B)}$

And so,

$$P(A \cap B) = P(A/B)P(B) \qquad \text{... (5)}$$

Now from (4) and (5) we get

$$P(B/A) = \frac{P(A/B)P(B)}{P(A)}$$

Which is (3) and this completes the proof.

8.5 GENERALIZED BAYES' THEOREM

If a sample space S can be divided into finitely many mutually exclusive events $B_1, B_2, \ldots B_n$ and if A is an event with nonzero probability which is a subset of $\bigcup_{i=1}^{n} B_i$, then for each B_i we have,

$$P(B_i/A) = \frac{P(A/B_i)P(B_i)}{P(A/B_1)P(B_1) + P(A/B_2)P(B_2) \ldots + P(A/B_n)P(B_n)}.$$

Proof of Generalized Bayes' Theorem

Since $A \subset \bigcup_{i=1}^{n} B_i$, we have

$$A = A \cap \left(\bigcup_{i=1}^{n} B_i \right)$$

$$= \bigcup_{i=1}^{n} (A \cap B_i).$$

Since $(A \cap B_i) \subset B_i$, $(i = 1, 2, ..., n)$ are mutually exclusive events, we have by addition law of probability

$$P(A) = P\left(\bigcup_{i=1}^{n} (A \cap B_i) \right) = \sum_{i=1}^{n} P(A \cap B_i)$$

$$= \sum_{i=1}^{n} P(B_i) P(A/B_i).$$

Also we have,

$$P(A \cap B_i) = P(A) P(B_i / A).$$

Therefore,

$$P(B_i / A) = \frac{P(A \cap B_i)}{P(A)}$$

$$= \frac{P(B_i) P(A/B_i)}{\sum_{i=1}^{n} P(B_i) P(A/B_i)}.$$

EXERCISES

1. If $P(B) = 0.5$, A and B are exclusive events, find

   (i) $P(A \cap B)$, (ii) $P(A \cap B')$, (iii) $P(A' \cap B)$, (iv) $P(A' \cup B')$,

   (v) $P(A) = 0.2$ $P(A \cup B)$.

2. A die is thrown four times. Getting a 1 or 6 is a success. Find the probability of (i) 4 successes, (ii) no success, (iii) at least one success, (iv) exactly 3 successes, (v) exactly 2 successes.

3. The probability that x solves the problem is 1/2 and y solves the problem is 2/3.

   What is the probability that,

   (i) x and y both solve the problem

   (ii) x solves and y fails

   (iii) x and y both fail to solve.

4. In a single throw of two die, find the probability of the following events:
   (i) odd number on the first die and 6 on the second
   (ii) a number > 4 on each die
   (iii) a total of 11
   (iv) a total > 8.

Solutions to Some Selected Problems on Probability

1. Five cards are drawn at random from a full pack. Find the probability that (a) they are of different denominations (b) two are of the same denominations and three scattered.

   ***Solution*:** 5 cards can be drawn from an ordinary pack of 52 cards in $^{52}C_5$ ways.

   (a) There are 13 denominations. The number of ways in which five different denominations can be chosen is $^{13}C_5$.

   Each of the five cards can be either of Clubs, Hearts, Spade or Diamond.

   So the number of favorable cases $= 13_{C_5} \times 4^5$.

   $$\therefore \text{ Probability of success} = \frac{13_{C_5} \times 4^5}{52_{C_5}} = 0.5070828.$$

   (b) Number of favorable cases = Number of ways of choosing two of same denomination, one each from different suits and three from different denominations out of the rest of 12 denominations,

   $= 13_{C_5} \times 4_{C_2} \times 12_{C_3} \times 4^3$.

   $$\therefore \text{ Probability of success} = \frac{13_{C_5} \times 4_{C_2} \times 12_{C_3}}{52_{C_5}} = 0.422569.$$

2. What is the probability that in a bridge game, (a) Each player gets a complete suit of cards, (b) at least one player has a complete suit of cards.

   ***Solution*:** Number of possible cases $= 52_{C_{13}} \times 39_{C_{13}} \times 26_{C_{13}} \times 13_{C_{13}}$

   $$= \frac{52!}{(13!)^4}$$

   (a) Since there are four suits and four people, the number of favorable, Cases = 4!

   $$\therefore \text{ probability of success} = \frac{4!}{\dfrac{52!}{(13!)^4}} = \frac{4! \times (13!)^4}{52!} = 4.47 \times 10^{-28}.$$

(b) Let A_i be the event that the ith player gets a complete suit. Probability that at least one of them gets a complete suit is the same as probability of $(A_1 \cup A_2 \cup A_3 \cup A_4) = P(A_1 \cup A_2 \cup A_3 \cup A_4)$. But, we know that for any 'n' events A_i, $i = 1, 2, \ldots, n$,

$$P\left(\bigcup_{i=1}^{n} A_i\right) = \sum_{i=1}^{n} P(A_i) - \sum_{i \neq j} P(A_i \cap A_j) + \sum_{i \neq j \neq k} P(A_i \cap A_j \cap A_k) \ldots.$$

Thus, $P(A_1 \cup A_2 \cup A_3 \cup A_4) = \sum_{i=1}^{4} P(A_i) - \sum_{i \neq j} P(A_i \cap A_j) +$

$$\sum_{i \neq j \neq k} P(A_i \cap A_j \cap A_k) - P(A_1 \cap A_2 \cap A_3 \cap A_4)$$

$$P(A_i) = \frac{4 \times 39C_{13} \times 26C_{13} \times 13C_{13}}{\frac{52!}{(13!)^4}} = \frac{4 \times 39! \times 13!}{52!}$$

$$\sum_{i=1}^{4} P(A_i) = \frac{4 \times 4 \times 39!}{52!}$$

$$P(A_i \cap A_j) = \frac{4_{C_2} \times 26_{C_{13}} \times 13_{C_{13}}}{\frac{52!}{(13!)^4}} = \frac{12 \times 26 \times (13!)^2}{52!}$$

$$P(A_i \cap A_j) = \frac{4_{C_2} \times 12 \times 26! \times 13!}{52!} = \frac{12 \times 26! \times (13!)^2 \times 6}{52!}$$

$$P(A_i \cap A_j \cap A_k) = \frac{4_{P_3 \times 13_{C_{13}}}}{\frac{52!}{(13!)^4}} = \frac{4 \times (13!)^4}{52!}$$

$$P(A_1 \cap A_2 \cap A_3 \cap A_4) = \frac{4! \times (13!)^4}{52!}$$

$$P(A_1 \cap A_2 \cap A_3 \cap A_4) = \frac{16 \times 39! \times (13!)}{52!} - \frac{16 \times 12 \times 26! \times (13!)^2}{52!} + \frac{3 \times 4! \times 3 \times (13!)^4}{52!}$$

probability of success = 2.52×10^{-1}

3. Cards are drawn from a pack without replacement. Find the probability that there are exactly 'k' cards between the first and second aces.

***Solution*:** Number of possible cases = 52!.

Number of ways in which the first ace comes at the 'x'th draw and the second ace comes at the $(k + x + 1)^{\text{th}}$ draw is

$$= 4_{C_2} \times 48_{P(k+x-1)} \times (51 - k - x)!$$

For the favorable case 'x' can vary from '1' to '$49 - k$.'

$$\therefore \text{ Number of favorable cases } = \sum_{x=1}^{49-k} {}^4P_2 \times {}^{48}P_{(k+x-1)} \times (51-k-x)$$

$$\sum_{x=1}^{49-k} \frac{4! \times 48! \times (51-k-x)!}{2! \times (49-k-x)!} = (48! \times 12) \sum_{x=1}^{49-k} (51-k-x)(50-k-x)$$

$\therefore$ probability of success.

$$p = \frac{48! \times 12}{52!} \sum_{x=1}^{49-k} (51-k-x)(50-k-x).$$

Let $50 - k - x = z.$

Then, $P = \dfrac{48! \times 12}{52!} \displaystyle\sum_{49-k}^{Z=1} (z+1)z$

$$\Rightarrow P = \frac{48! \times 12}{52!} \frac{(49-k)(50-k)(51-k)}{3}$$

$$= \frac{48! \times 4 \times (51-k)(50-k)(49-k)}{52!} = \frac{{}^{51-k}C_3}{{}^{52}C_4}.$$

4. Fifteen students among whom are A and B, stand in a (a) row, (b) ring. What is the probability that there are exactly five students between A and B?

   (a) Total number of possible cases = 15!

   number of ways in which one of A or B stands in the xth place and the other in $(x+6)^{\text{th}}$ place = 2 × 13!

   For a favorable case 'x' can vary from '1' to '9'.
   Thus the probability of success

$$= \frac{\sum_{x=1}^{9}(2 \times 13!)}{15!} = \frac{2 \times 9 \times 13!}{15!} = 0.0857142.$$

   (b) Total number of possible cases = 14!

   Total number of favorable cases = 2 × 13!

   (We fix a ring by fixing a particular man and then clockwise or anticlockwise of the particular man people stand for different positions i.e. total number of cases = 14!
   Further if A's position is the fixed position of the ring, B can stand.

   Keeping 5 men between them only in '2' ways one clockwise and the other anticlockwise and again all other '13' people can stand in 13! ways).

$$\therefore \text{ Probability of success} = \frac{2\times 13\,!}{14!} = \frac{2}{14} = \frac{1}{7}$$

5. A man writes 'n' cheques and puts them at random in 'n' addressed envelopes. What is the probability that not a single cheque goes into its appropriate envelope?

Solution: Let A_i be the event that i^{th} cheque goes to its proper envelope. Therefore, at least one cheque goes to its appropriate envelope, is $\bigcup_{i=1}^{n} A_i = S^c$ (where S denotes the event of success)

$$P(S^C) = P(\bigcup_{i=1}^{n} A_i) = P(A_i) - P(A_i \cup A_j) + \ldots + (-1)^{n+1} P(A_1 \cap A_2 \cap \ldots \cap A_n)$$

$$P(A_i) = n \times \frac{1}{n} = 1$$

$$P(A_i \cap A_j) = \frac{1}{n}\frac{1}{n-1} = \frac{1}{n(n-1)}$$

$$\Rightarrow P(A_i \cap A_j) = n_{C_2} \frac{1}{n(n-1)}$$

$$P(A_i \cap A_j \ldots \cap A_m) = \frac{1}{n}\frac{1}{n-1}\frac{1}{n-2}\cdots\frac{1}{n-k+1}$$

$$\Rightarrow \sum P(A_i \cap A_j \ldots A_m) = n_{C_k} \frac{1}{n(n-1)(n-k+1)}$$

$$P(A_i \cap A_j \cap A_k \ldots A_n) = \frac{1}{\text{n!}}$$

$$P(S^c) = n \times \frac{1}{n} - n_{C_2} \frac{1}{n(n-1)} \ldots (-1)^{n-1} n_{C_n} \frac{1}{n!}$$

$\Rightarrow$ The required probability $= P(S) = 1 - P(S^C)$

$$P(S) = 1 - 1 - \frac{1}{2!} + \frac{1}{3!} \ldots (-1)^{n-1} \frac{1}{n!} = \sum_{k=0}^{n} \frac{(-1)^k}{k!}$$

6. Suppose $P(A)$, $P(B) > 0$ (a) If A and B are mutually exclusive will they be independent? (b) If A and B are independent, can they be mutually exclusive?

Solution:

(a) A and B are mutually exclusive

$\Rightarrow$ A prevents the occurrence of B (or B prevents the occurrence of A)

$\Rightarrow$ A and B are independent.

(b) A and B are independent $\Rightarrow P(A \cap B) = P(A).P(B)$;
$P(A), P(B > 0) \Rightarrow P(A \cap B) > 0 \Rightarrow$ A and B cannot be mutually exclusive.

7. It is 9 to 5 against a person who is 50 years living till he is 70 and 8 to 6 against a person who is 60 living till he is 80. Find the probability that (a) only one of the two persons survive after 20 years. (b) At least one of the two persons survive after 20 years.

(a) Probability of only one of the two persons survive after 20 years = P (one will die in 50–70) × P (one will survive in 60–70) + P (one will die in 60–80) × P (one will survive in 50–70),

$$= \frac{9}{14} \times \frac{6}{14} \times \frac{5}{14} \times \frac{8}{14} = 0.4795918$$

(b) Probability that both die in 20 years $= \frac{9}{14} \times \frac{8}{14}$

$\Rightarrow$ Probability that at least one survives $= 1 - \frac{72}{196}$

$$= \frac{124}{196} = 0.63265$$

8. Suppose a biased coin has probability 'P' of scoring head. Find the probability that 'a' heads will come before 'b' tails.

Solution: Probability that exactly 'x' tails occur before 'a' heads are completed,

$= {a+x-1}_{C_{a-1}} p^{(a-1)} (1-p)^x . p$

For the required event, x can vary from '0' to 'b–1'.

$\therefore$ The required probability $= \sum_{x=0}^{b-1} {a+x-1}_{C_{a-1}} p^a (1-p)^x$

9. A and B toss 'n' and 'n + 1' coins (all unbiased). Find the probability that,

(a) A gets 'x' heads in 'n' tosses

$= n_{C_x} \left(\frac{1}{2}\right)^x \left(\frac{1}{2}\right)^{n-x}$

(b) B gets 'x' heads in 'n' tosses

$= {n+1}_{C_x} \left(\frac{1}{2}\right)^x \left(\frac{1}{2}\right)^{n+1-x}$

both A and B have 'x' heads

$= n_{C_x} \, {n+1}_{C_x} \left(\frac{1}{2}\right)^{2x+1}$

both A and B have same number of heads

$$= \sum_{x=0}^{n} n_{C_x} n+1_{C_x} \left(\frac{1}{2}\right)^{2x+1}$$

(c) A gets $'x+\alpha'$ heads in 'n' tosses

$$= n_{C_{x+\alpha}} \left(\frac{1}{2}\right)^{n}$$

$\Rightarrow$ A gets more heads than B

$$= \sum_{x=0}^{n-1} \sum_{\alpha=1}^{n-x} n_{C_{x+\alpha}} n+1_{C_x} \left(\frac{1}{2}\right)^{2n+1}$$

(d) A gets '$x - a$' heads in 'n' tosses

$$= n_{C_{x-\alpha}} \left(\frac{1}{2}\right)^{n}$$

$\Rightarrow$ A gets less heads than B

$$= \sum_{x=1}^{n+1} \sum_{\alpha=1}^{x} n_{C_{x-\alpha}} n+1_{C_x} \left(\frac{1}{2}\right)^{2n+1}$$

10. A man has 'n' keys exactly one of which opens a lock. If he tries the keys at random, what is the probability that he opens the lock at the 'k' th attempt if (a) unsuccessful keys are kept aside (b) they are not.

***Solution*:**

(a) Probability of success at the first trial $= \frac{1}{n}$

Probability of success at the second trial $= \frac{1}{n-1}$

⋮

Probability of success at the 'k' trial $= \frac{1}{n-(k-1)}$.

Therefore probability of first success at the second trial = (Probability of failure in the first trial) × (Conditional probability of success at the second trial),

$$= \left(1 - \frac{1}{n}\right)\frac{1}{n-1} = \frac{1}{n}.$$

Similarly probability of first success at the third trial,

$$= \left(1 - \frac{1}{n}\right)\left(1 - \frac{1}{n-1}\right)\frac{1}{n-2}$$

$$= \frac{1}{n}.$$

So, probability of first success at the 'k' trial,

$$=\left(1-\frac{1}{n}\right)\left(1-\frac{1}{n-1}\right)\ldots\left(1-\frac{1}{n-k}\right)\times\frac{1}{n-k+1}=\frac{1}{n}$$

(b) Probability of success at the first trial $=\frac{1}{n}$. Probability of failure at the first trial $=1-\frac{1}{n}$ and is same for all trials since the keys are not eliminated.

Therefore, probability of first success at the 'k' trial $=\left(1-\frac{1}{n}\right)^{k-1}\times\frac{1}{n}$.

11. In a bolt factory, machines A, B and C manufacture respectively 25, 5 and 40 percent of the total of their output 5, 4 and 2 percent are defective. The output of the three machines are kept in three boxes. One box is chosen at random and a bolt is taken from it. It is found to be defective.

What is the probability that the box contained bolts produced by A? by B? by C? If another bolt is taken from the same box, what is the probability that this is also defective?

Solution: Let A be the box containing output of machine A, B be the box containing output of machine B and C be the box containing output of machine C.

Probability that any one of the boxes chosen at random

$$=P(A)=P(B)=P(C)=\frac{1}{3}.$$

Let D be the output found to be defective

$$\therefore\ P(D/A)=0.05,\ P(D/B)=0.04,\ P(D/C)=0.02$$

$$\Rightarrow P(A/D)=\frac{P(A)P(D/A)}{P(A)P(D/A)+P(B)P(D/B)+P(C)P(D/C)}$$

$$=\frac{0.05}{0.05+0.04+0.02}=\frac{5}{11}$$

$$P(B/D)=\frac{0.04}{0.11}=\frac{4}{11}$$

$$P(C/D)=\frac{0.02}{0.11}=\frac{2}{11}.$$

Probability of another defective from the same box

$= P(A/D) \times 0.05 + P(B/D) \times 0.04 + P(C/D)0.02$

$= \frac{5}{11} \times \frac{5}{100} + \frac{4}{11} \times \frac{4}{100} + \frac{2}{11} \times \frac{2}{100} = \frac{45}{1100} = 0.0409.$

12. A die is thrown as long as necessary for an ace to turn up. Assuming that the ace does not turn up at the first throw what is the probability that more than three throws will be necessary?

Solution: Let A be the event that ace does not turn up at the first throw and B be the event that more than 3 throws will be necessary.

$$P(B/A) = \frac{P(B \cap A)}{P(A)} = \frac{P(B)}{P(A)} \text{ (Since } B \text{ is a subset of } A)$$

$$P(B) = \sum_{x=4}^{\infty} \left(1 - \frac{1}{6}\right)^{x-1} \left(\frac{1}{6}\right) = \frac{\left(1 - \frac{1}{6}\right)^3}{1 - \left(1 - \frac{1}{6}\right)} \left(\frac{1}{6}\right) = \left(1 - \frac{1}{6}\right)^3$$

$$P(A) = \sum_{x=2}^{\infty} \left(1 - \frac{1}{6}\right)^{x-1} \left(\frac{1}{6}\right) = \frac{\left(1 - \frac{1}{6}\right)}{1 - \left(1 - \frac{1}{6}\right)} \left(\frac{1}{6}\right) = \left(1 - \frac{1}{6}\right)$$

$$\Rightarrow P(B/A) = \frac{\left(1 - \frac{1}{6}\right)^3}{1 - \frac{1}{6}} = \left(1 - \frac{1}{6}\right)^2 = \frac{25}{36}$$

13. Show that $P(A/B) = P(A) \Rightarrow P(B/A) = P(B)$ if $P(B) \neq 0$.

Solution: $P(A/B) = \frac{P(A \cap B)}{P(B)} = P(A)$

$$P(B/A) = \frac{P(A \cap B)}{P(A)} = P(A).\frac{P(B)}{P(A)} = P(B).$$

CHAPTER 9

Theory of Computation

9.1 FINITE STATE MACHINE

In this chapter we shall discuss the basic ideas of Finite State Machine or what is called finite Automata. Since computer science students have a separate course on this subject, usually they do not study this topic under Discrete Mathematics course. But this topic forms a part of this course in most of the Universities for mathematics students, and hence we include this chapter mostly for the benefit of such students.

This topic is used in spell checking, grammar checking, used as speech and language recognizer. This is also used to model a vending machine where coins are given as input and required item, if available is obtained as output.

We now start with a concept of what is called a computing device. It consists of a computing machine and programs. We shall first discuss some examples.

Basic Machine

A basic machine is a computing device which recognizes an input and produces an output.

The simplest example of a basic machine is the load and the output is the weight printed card. As another simple example we may consider the multiplications table such as: $8 \times 7 \Rightarrow 56$.

Thus a basic machine can be mathematically defined precisely as a function (mapping),

$$F: I \to O$$

where I is the input and O is the output.

Finite State Machine

A machine can have some sort of internal memory that can keep track of certain information about inputs. A complete internal condition of the machine along with all its memory at any point of time is said to constitute the state of the machine at that time.

In fact the input set I and state set S are finite sets in case of finite state machines. As an example, consider

$$237(6 \times 7)\sigma_0 \longrightarrow 4 \equiv F_1 : I \times S \longrightarrow O$$
$$\times\, 576$$

$$\text{2 carry out } 4\ (6 \times 7)\sigma_0 \longrightarrow 4 \equiv F_2 : I \times S \longrightarrow S$$

In this case the machine consists of S, I and two functions F_1 and F_2.

We are now in a position to quote precise definition.

A finite state machine is defined as a $(S, I, \mathfrak{F})$ where S and I are finite sets respectively called the state set, input set and the third one is a finite family of functions.

A variety of extra features can be added in a finite state machine to further increase the utility. One useful extension is what is called a Moore machine or a recognition machine, M. This is defined as$(S, I, \mathfrak{F}, s_0, T)$ where $s_0 \in T$ and $T \subseteq S$. The state s_0 is called the starting state of acceptable states of M. These states are used in connection with language recognition.

The theory of computation deals with the techniques to **assess** the power of efficiency of computing device.

There are certain limitations of usual finite state machine and it cannot be used to any general model. Therefore one has to go for Turing Machine. Turing machine was invented by Alan Mathison Turing while trying to solve the decision problem posed by Hilbert. The problem was whether there exists a general method which can be applied to any assertion to determine whether the assertion is true.

We shall not attempt to discuss Turing machine in this introductory book, it is out of the scope of the book, and out of the syllabus in the first course of Discrete mathematics for UG Mathematics students.

With this introductory remark we now come to the topic. We start with quoting the definitions of string, word, alphabet and language etc.

Symbol: Symbols are undefined concepts or primitives. Examples: $a, b, c, \cdots$.

String or word: Any sequence of symbols (or any succession of symbols) juxtaposed (written one after the other) is called a string.

9.1.1 ***Example:***$\Sigma = \{a, b\}$
Strings or words: $a, b, ab, ba, aab, bba, aaa, \cdots$.

A word without a letter is called empty word and is written as $\wedge$ (capital Gamma).

Let $\sum$ denote the set of words
and $\sum^*$ the set of all words over $\sum$.

Length of a string or word: Let $x \in \sum^*$, then length of x, denoted by $|x|$ is the number of letters or symbols in it. Length of a string is the number of symbols composing the string. Thus

$$|\wedge| = 0, |a| = 1, |aaa| = 3, \cdots$$

Concatenation of two strings: The concatenation of two strings is the string formed by writing the first followed by the second with no intervening space.

9.1.2 ***Example:*** Dog, house → doghouse

9.1.3 ***Example:*** Let $\sum = \{a, b\}$.

Let $U \subset \sum^*$ be such that $U = aba$ is a word.

Let $V \subset \sum^*$ be such that $V = bba$ is a word.

Then concatenation of U and V, is

$UV = aba\ bba$ is a word.

Similarly if $U = aa, V = bb$, then

$$UV = aabb.$$

Alphabet: An alphabet is a finite set of symbols. Let A be a finite set which is the alphabet of the language. Let $A^* = \{a, b, c, \cdots, x, y, z\}^*$be the set of all words (the set of all sequences of symbols) over A, i.e., all sequences of one letter, two letters, three letters etc.

Language: A language (over the alphabet A) is a subset of A^*.

Example, $A = \{a, b, c\}$.

The following are the languages over A.

$$L_1 = \{a, aa, ab, ac, abc, cab\}$$
$$L_2 = \{aba, aabaa\}$$
$$L_3 = \{a^i cb^i, i \geq 1\}.$$

In other words, a language (over the alphabet A) is a set of strings of symbols from the alphabet A.

Operation on the set of words

(i) *Union:* Let $U, V \in \Sigma^*$. Then
$$U \cup V = \{w \in \Sigma^* : w \in U \text{ or } w \in V\}.$$

(ii) *Product or concatenation:*
Let $U, V \in \Sigma^*$,
$UV = \{uv \in \Sigma^* : u \in U, v \in V\}$.

9.1.4 ***Example:*** $\Sigma = \{a, b\}, U = \{aa, bb\}, V = \{ab, ba\}$.
$UV = \{aaab, aaba, bbab, bbba\}$.

Remarks:

(a) Concatenation is associative.
Let$U, V \in \Sigma^*$. Then
$(UV)W = U(VW)$.
Let $x \in (UV)W \Leftrightarrow x = (uv)w$ for $u \in U, v \in V, w \in W$
$= uvw = u(vw)$.
Hence, $(UV)W = U(VW)$.

(b) Concatenation is not commutative.
$UV \neq VU$.
Let $U = \{a\} \subset \Sigma^*, V = \{b\} \subset \Sigma^*$.
$UV = \{ab\} \neq \{ba\} = VU$.

(c) *Star or Closure Operation:* Let $S \subset \Sigma^*, \Sigma^*$, the closure set of *S*, is the set consisting of $\wedge$ (empty word) and all finite concatenation of words belonging to *S*.

9.1.5 ***Example:*** $S = \{ab, ba\}$.

$$S^* = \{\wedge, ab, ba, abab, abba, baba, baab, P, Q, R\}$$

where$P = ababab, Q = ababba$,
$R = abbaba$ etc.

Regular sets: A class $\mathscr{k}$ of subsets of Σ^* is called a regular set if

(i) $\{\wedge\}, \emptyset \in \mathscr{k}$,

(ii) all finite set of words (all finite subsets of Σ^*) $\in \mathscr{k}$,

(iii) $R_1, R_2 \in \mathscr{k} \Rightarrow R_1 \cup R_2 \in \mathscr{k}, R_1R_2 \in \mathscr{k}$,
$R \in \mathscr{k} \Rightarrow R^* \in \mathscr{k}$.

9.1.6 ***Example:*** $\Sigma = \{ab\}$.
The class of all sets of words having two consecutive a's is a regular set.

Suppose $\mathscr{k} = \{\{\emptyset\}, \{a, a\}, \{a, a, b\}, \{b, a, a\}, \{ba, ab\} \cdots\}$,

$$R_1 = \{b, aa\}, R_2 = \{a, ab\}.$$

$$R_1 \cup R_2 = \{baa, aab\} \in \mathscr{k}$$

$$R_1R_2 = \{baa, aab\} \in \mathscr{k}$$

R is regular set.

Nonregular set: A class $\mathscr{k}$ of the subsets of Σ^* is called a nonregular set if it does not obey the axioms of a regular set.

9.1.7 ***Example:*** Suppose $\mathscr{k} = \{\{\emptyset\}, \{a, a\}, \{a^2b^2\}, \{a^3b^3\}, \{a^4, b^4\}, \cdots\}$,

If $R_1 = \{a, b\}, R_2 = \{a^2, b^2\}$.
$R_1 \cup R_2 \in \mathscr{k}$ but $R_1R_2 \notin \mathscr{k}$
Since $R_1R_2 = \{aa^2, ab^2, ba^2, b^3\}$.

9.2 FINITE AUTOMATION

A finite automation or finite state machine say A on $\Sigma = \{\sigma_1, \sigma_2, \cdots, \sigma_n\}$ is a finite directed graph in which, from each vertex lead out 'n' arrows labelled by distinct $\sigma_i (1 \leq i \leq n)$. There is one vertex labelled by – sign, called the initial vertex and a set of vertices labelled by + sign, called the final vertices. Initial vertex may be sometimes same as the final vertex.

The vertices are called the states and the finite automation is called a finite state machine or is said to represent a finite state machine.

9.2.1 ***Example:*** $\Sigma = \{a, b\}$

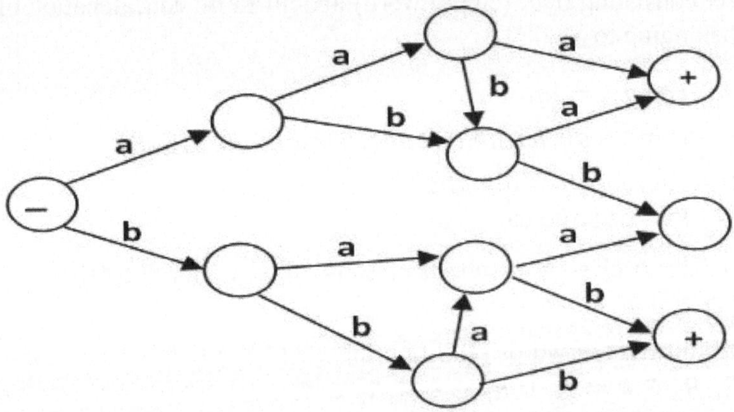

9.2.2 ***Example:*** Dividing by 3 tester machine.

Input: $I = \{0, 1, 2, 3, 4, 5, 6, 7, 8, 9\}$
$S = \{-, s_0, s_1, s_2\}$ where – is the initial vertex
$\sigma_0 = \{0, 3, 6, 9\}, \sigma_1 = \{1, 4, 7\}, \sigma_2 = \{2, 5, 8\}$,
$\Sigma = \{\sigma_0, \sigma_1, \sigma_2\}$

W-path: A path in a finite automation, from a vertex i to a vertex j is called a W-path if the concentration is of labels the path from the word.

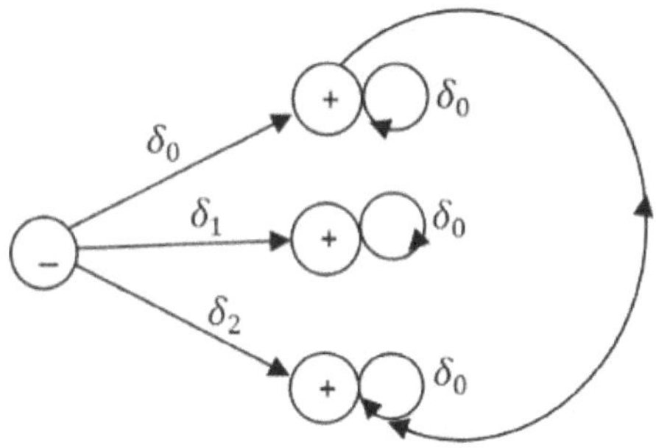

9.2.3 ***Example:*** $\Sigma = \{a, b\}$

A word is said to be accepted by a finite automation, say *A,* if there is a W-path from the initial vertex to a final vertex.

9.3 SUM OVER SETS OF WORDS

Let R_1 and R_2 be words accepted by an automation (graph). Then $R_1 + R_2$ is the alternate path of R_1 and R_2 such that,

$$R_1 + R_2 = R_1 \cup R_2.$$

Regular Expression

Let Σ be the set of alphabets. A regular expression on Σ is defined recursively as follows:

1. Λ and Φ, empty word, and set of empty words is regular expressions
2. Every letter $\sigma \in \Sigma$ is a regular expression
3. R_1 and R_2 are regular expressions
   $\Rightarrow R_1 + R_2$ and R_1R_2 are regular expressions.
4. R is regular expression
   $\Rightarrow R^*$ is regular expression.
5. Nothing else is regular expression (*re*) that is not obtained by rules 1–4.

9.3.1 ***Example:*** $\Sigma = \{\text{a}, \text{b}\}$.

1. a is a letter and hence is a *re.*
2. $a + b$ is the sum of two *re* and hence is a regular expression
3. a^*, b^* are regular expressions
4. $a + b.a^*$ is a *re*
5. $(a + b.a^*)ab^*$ is a *re.*

If $\Sigma = \{\text{a}, \text{b}\}$, then $(a +. b), (a. b)$ are not *re.*

Set Associated with a Regular Expression

For a regular expression R, the set usually denoted by $\tilde{R}$ (pronounced as R tilta) is defined recursively as follows:

(i) $R = \Lambda \Rightarrow \tilde{R} = \{\Lambda\}$

$R = \emptyset \Rightarrow \tilde{R} = \{\emptyset\}$

$R = \sigma \Rightarrow \tilde{R} = \{\sigma\}$ for $\sigma \in \Sigma$

(ii) $R = R_1 + R_2 \Rightarrow \tilde{R} = \tilde{R}_1 \cup \tilde{R}_2$

$R = R_1R_2 \Rightarrow \tilde{R} = \tilde{R}_1\tilde{R}_2$

(iii) $R = R_1^* \Rightarrow \tilde{R} = \left(\tilde{R}_1\right)^*$

R, S are re such that $R = S \Rightarrow \tilde{R} = \tilde{S}$.

9.4 TRANSITION GRAPH

A graph T is called a transition graph on Σ iff T is a finite directed graph in which arrow is labelled by word $W \in \Sigma^*$ and in which there is a set of vertices labelled by '$-$' sign (called the initial vertices) and a set (possibly empty) of vertices labelled by '$+$' sign called the final vertices.

Initial vertex can also be a final vertex.

9.4.1 *Example:* $\Sigma = \{a, b\}$.

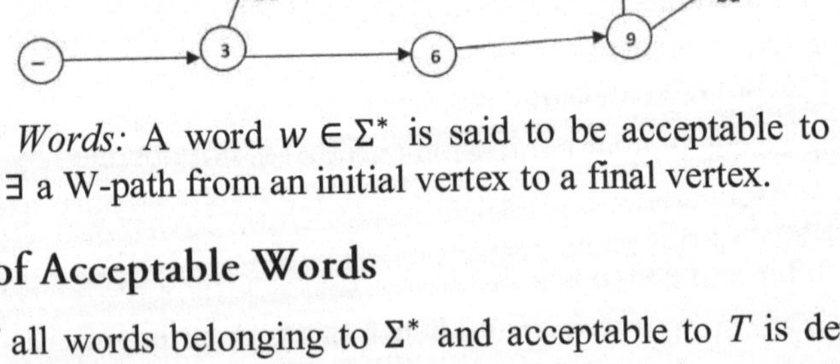

Acceptable Words: A word $w \in \Sigma^*$ is said to be acceptable to a transition graph T iff $\exists$ a W-path from an initial vertex to a final vertex.

The Set of Acceptable Words

The set of all words belonging to Σ^* and acceptable to T is denoted by $\tilde{T}$. Then

$$\tilde{T} = \{w \in \Sigma^* : w \text{ is acceptable to } T\}.$$

9.4.2 *Example:* $\Sigma = \{\text{a, b}\}$.

$\tilde{T} = \{\Lambda, aabb, bb\}$

$\tilde{T} = \{ab, bbba, bb\}$.

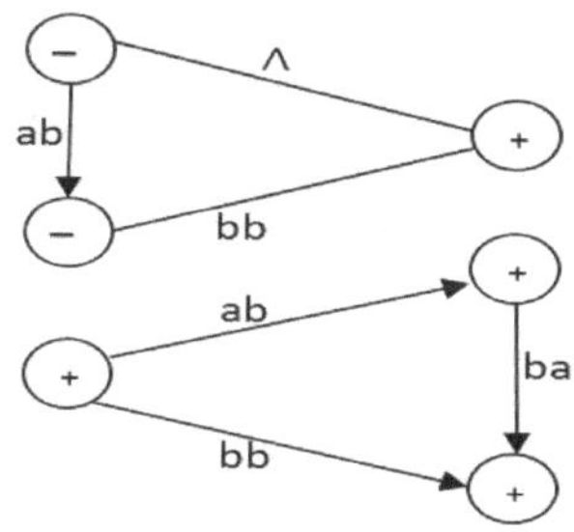

9.5 GENERALIZED TRANSITION GRAPH

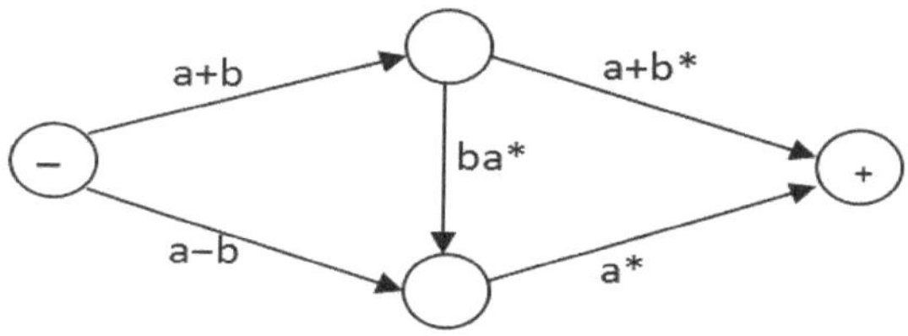

Generalized transition graph is a finite directed graph in which each arrow is labelled by a regular expression.

9.5.1 ***Example:*** $\Sigma = \{\text{a}, \text{b}\}$ where $(a+b), (a+b^*), a^*, a-b$ are not words but regular expressions.

The identifies on Regular Expressions

I. (i) $R \wedge = \wedge R = R$

(ii) $\wedge + R = R + \wedge = R$

(iii) $\emptyset R = R\emptyset = R$

(iv) $\emptyset + R = R + \emptyset = R$

(v) $R + R = R$

II. (i) $R + S = S + R$

(ii) $RS \neq SR$

III. (i) $(R + S) + T = R + (S + T)$

(ii) $R(ST) = (RS)T$

IV. (i) $R(S + T) = RS + RT$

(ii) $(S + T)R = SR + TR$

V. (i) $R^* = RR^* = R^*R$

$= R^*R^* = (R^*)^*$

(ii) $R^* = \wedge + RR^*$

VI. (i) $(RS)^*R = R(SR)^*$

(ii) $(R+S)^* \neq R^* + S^*$

$(R+S)^* = (R^*S^*)^* = (R^* + S^*)^*$

Proof:

I. (i) $R\wedge = \wedge R = R$

Let $r \in R$. We know $\wedge r = r \Rightarrow \wedge r \in R$.

By the property of concatenation

$\wedge R = \{\lambda r: \lambda \in \wedge, r \in R\}$.

Hence, $\wedge R = R$. Similarly it can be proved that $R\wedge = R$.

(ii) $\wedge + R = R + \wedge = R$

Let $s \in \wedge + R$.

Hence

$$x = \lambda + r = r \in R.$$

So $R = \wedge + R$. Similarly, let $y \in R + \wedge$.

Then

$$y = r + \lambda \text{ for } r \in R$$

$$\Rightarrow y \in R \text{ (since } r + \lambda = r\text{). Hence } R = R + \wedge.$$

Proof of (iii) is trivial.

(iv) $\emptyset + R = R + \emptyset = R$

Let $x \in \emptyset + R \Leftrightarrow x = \emptyset + r = r \Leftrightarrow \emptyset + R = R$.

Similarly $y \in R + \emptyset \Leftrightarrow y = r + \emptyset = r$

$$\Leftrightarrow R = R + \emptyset.$$

(v) $R + R = R$.

$x \in R + R \Leftrightarrow x = r + r, r \in R$

$\Leftrightarrow x = \{r\} \cup \{r\} = \{r\} = r$

$\Leftrightarrow R + R = R$.

II. (i) $R + S = S + R$

$x \in R + S \Rightarrow x = r + s$ for $r \in R, s \in S$

$\Leftrightarrow x = \{r\} \cup \{s\} = \{s\} \cup \{r\}$

$= s + r$

$\Leftrightarrow x \in S + R$

$\Rightarrow R + S = S + R$.

(ii) $RS \neq SR$

Let $R \ni a, S \ni b$. Then $RS \ni ab, SR \ni ba$.

But $ab \neq ba$, hence $SR \neq RS$.

III. (i) $(R+S)+T = R+(S+T)$

$x \in (R+S)+T \Leftrightarrow x = (r+s)+t, r \in R, s \in S, t \in T$

$\Leftrightarrow\ x = (r+s+t)$

$\Leftrightarrow\ x = r+(s+t)$

$\Leftrightarrow\ x \in R+(S+T)$

(ii) $R(ST) = (RS)T$

Let $x \in R(ST)$ for $R \ni a, S \ni b, T \ni c$

$x = a(bc) = (abc) = (ab)c$

$\Leftrightarrow\ x \in (RS)T.$

IV. (i) $R(S+T) = RS+RT$

$x \in R(S+T) \Rightarrow x \in r(s+t)$ where $r \in R, s \in S, t \in T$

$\Leftrightarrow\ x = rs+rt \in RS+RT$

$\Leftrightarrow\ R(S+T) \subset RS+RT$

Similarly$y \in RS+RT$

$\Leftrightarrow\ y = rs+rt = r(s+t) \in R(S+T)$

$\Leftrightarrow\ RS+RT \subset R(S+T)$

Hence

$RS+RT = R(S+T)$

(ii) $(S+T)R = SR+TR$

$x \in (S+T)R \Rightarrow x \in (s+t)r$ where $r \in R, s \in S, t \in T$

$\Leftrightarrow\ x = sr+tr \in SR+TR$

$\Leftrightarrow\ (S+T)R \subset SR+TR$

$y \in SR+TR \Rightarrow y = sr+tr$

$\Leftrightarrow\ y = (s+t)r \in (S+T)R$

$\Leftrightarrow\ SR+TR \subset (S+T)R.$

Hence

$SR+TR = (S+T)R.$

V. (i) $R^* = RR^* = R^*R = R^*R^* = (R^*)^*$

$R^* = RR^*$

By definition of R^*, we have

$R^* = R^0 \cup R^1 \cup R^2 \cup \cdots \cup R^k.$

Let $x \in R^* \Rightarrow x = r_1r_2r_3 \cdots r_k$where $r_i \in R^i$

$\Rightarrow\ x \in r_1(r_2r_3 \cdots r_k) \in RR^*$

$\Rightarrow\ R^* \subset RR^*$

Let $y \in RR^* \Rightarrow y = r_1(r_2r_3 \cdots r_k) = r_1r_2r_3 \cdots r_k \in R^*$

$\Rightarrow\ RR^* \subset R^*$

Thus $R^* = RR^*$

$R^* = RR^*$
$x \in R^* \Rightarrow x = r_1 r_2 r_3 \cdots r_k = (r_1 r_2 r_3 \cdots r_{k-1}) r_k \in R^* R$
$\Rightarrow R^* = R^* R$
$(R^*)^* = R^*$
$x \in (R^*)^* \Leftrightarrow x = r_1^* r_2^* r_3^* \cdots r_k^*$
where $r_1^* \in R^*$
$r_i^* \in R^* \Rightarrow r_i^* = r_{i_1} r_{i_2} \cdots r_{i_k}$
$x \in (R^*)^* \Rightarrow x = r_{1_1} r_{1_2} \cdots r_{1_k} r_{2_1} r_{2_2} \cdots r_{2_k} \cdots r_{k_1} r_{k_2} \cdots r_{k_k} \in R^* R$
$\Rightarrow x \in R^* \Rightarrow (R^*)^* \subset R^*$
Reversing the above process, $R^* \subset (R^*)^*$
Hence $(R^*)^* = R^*$.

(ii) $R^* = \wedge + RR^*$
We know$R = \wedge + R \Rightarrow \wedge + R^* = \wedge R^* R$ [$\because R = R^* R$]
$\Rightarrow R^* = \wedge + R^* = \wedge + (R^*)^* R^*$
$\Rightarrow \wedge + R^* R^* = \wedge RR^*$
$\Rightarrow R^* = \wedge + RR^*$

9.5.2 ***Example:*** Prove that the set of strings in which every 0 is immediately followed by at least two 1's can be described by both R_1 and R_2 where

$$R_1 = \wedge + 1^*(011)^*(1^*(011)^*)^*$$
$$R_2 = (1 + 011)^*$$

Proof: $R_1 = \wedge + 1^*(011)^*(1^*(011)^*)^*$

$= (1^*(011)^*)^*$ since $\wedge + RR^* = R^*$

$= (1 + (011)^*)^*$ since $(R^* S^*)^* = (R + S)^*$

$= R_2$

9.5.3 ***Example:*** Prove the identity

$$(1 + 00^*1) + (1 + 00^*1)(0 + 10^*1)^*(1 + 10^*1)$$
$$= 0^*1(1 + 10^*1)^*$$

Proof:

$$(1 + 00^*1) + (1 + 00^*1)(0 + 10^*1)^*(0 + 10^*1)$$
$$= (1 + 00^*1)[\wedge + (0 + 10^*1)^*(0 + 10^*1)]$$
$$= [(\wedge + 00^*)1][\wedge + (0 + 10^*1)^*(0 + 10^*1)]$$
$$= 0^*1(1 + 10^*1)^*$$

Theorem (Arden's rule) Let $\wedge \notin S$. Then

(i) $R = SR + T \Leftrightarrow R = S^*T$

(ii) $R = RS + T \Leftrightarrow R = TS^*$

where R, S and T are regular expressions on Σ.

Proof:

(i) $\Leftarrow: R = S^*T \Rightarrow R = SR + T$

Now$R = S^*T \Leftrightarrow R = (\wedge + S^*)T = \wedge T + S^*T$

$\Leftrightarrow R = \wedge T + (SS^*)T$ (since $R^* = RR^*$ and $\wedge + R = R$)

$= T + S(S^*T)$

$= T + SR$ since $R + S = S + R$.

In order to prove $(:\Longrightarrow)$ part we need the following lemma.

Lemma$R = SR + T \Rightarrow$

$R = S^{l+1}R + (S^lR + S^{l-1}R + \cdots + SR + T)$

for any positive integer l.

Proof of the Lemma. We have

$R = SR + T = S(SR + T) + T$ (since $R = SR + T$)

$\Rightarrow R = S^2R + ST + T$

$\Rightarrow R = S^2(SR + T) + ST + T = S^3R + S^2T + ST + T$

Repeating this process l times we have

$R = S^{l+1}R + S^lT + S^{l-1}T + \cdots + ST + T$

$= S^{l+1} + \left(S^lT + \cdots + ST + T\right).$

Proof of: $\Rightarrow$

$R = SR + T$. Let $x \in R$ and let the length of x, $|x| = l$.

By the lemma,

$R = S^{l+1}R + \left(S^lT + S^{l-1}T + \cdots + ST + T\right).$

For any $w \in S^{l+1}R$, and $\wedge \notin S$,

$|w| \geq l + 1$.

Hence, $s \notin S^{l+1}R$ (since $|x| = l$, but $S^{l+1}R$contains words of lengths $\geq l + 1$)

$\Rightarrow x \in S^lT, 0 \leq i \leq l$

$\Rightarrow x \in S^*T$ (since $S^l \subset S^*$)

$\Rightarrow R \subset S^*T.$

We have to show that $S^*T \subset R$.

Let$x \in S^*T$. But $S^* = S^0 \cup S^1 \cup \cdots \cup S^l \cup \cdots$.

Therefore $x \in S^kT$ for $k = 0, 1, 2, \cdots$.

By the lemma, R can be expressed

$R = S^{k+1}R + (S^kT + \cdots + ST + T)$

$\Rightarrow S^kT \subset R \Rightarrow x \in R$.

Hence, $S^*T \subset R$ and thus $S^*T = R$.

(ii) $R = TS^* \Rightarrow R = RS + T$

Now $R = TS^* \Rightarrow$

$R = T(S^* + \wedge) = T(S^*S) + T\wedge$

$= (TS^*)S + T\wedge = RS + T$

$: \Rightarrow R = RS + T \Rightarrow R = TS^*$.

Let $x \in R, |x| = l$.

By the lemma

$R = RS^{k+1} + (TS^k + TS^{k-1} + \cdots + ST + T)$

Let $w \in RS^{l+1} \Rightarrow |w| \geq l + 1$.

$\Rightarrow x \notin RS^{k+1}$

$\Rightarrow x = (TS^l + TS^{l-1} + \cdots + TS + T)$

$\Rightarrow x \in TS^i, 0 \leq i \leq 1$

$\Rightarrow x \in TS^* \; (S^i \subset S^*)$

Hence $R \subset TS^*$.

Let $x \in TS^*$. But

$S^* = S^0 \cup S^1 \cup \cdots \cup S^i \cup$

Hence, $x \in TS^k$ for some $k = 0, 1, 2, \cdots$.

By the lemma,

$R = RS^{k+1} + TS^k + TS^{k-1} + \cdots + TS + T)$

$\Rightarrow TS^k \subset R$

$\Rightarrow TS^* \subset R$

Thus $TS^* = R$.

Application: Prove that,

$$(b + aa^*b) + (b + aa^*b)(a + ba^*b)^*(a + ba^*b)$$
$$= a^*b(a + ba^*b)^*$$

Proof: $LHS = (b + aa^*b) + (b + aa^*b)(a + ba^*b)^*(a + ba^*b)$
$= (b + aa^*b) \wedge +(b + aa^*b)(a + ba^*b)^*(a + ba^*b)$(By I)
$= (b + aa^*b)(\wedge +(a + ba^*b)^*(a + ba^*b))$(By IV)
$= (b + aa^*b)(\wedge +(a + ba^*b)^*)$ since $(a + ba^*b)^*(a + ba^*b)$
$= (a + ba^*b)^*$
$= (b + aa^*b)(a + ba^*b)^*$ since $\wedge +R = R$
$= (\wedge b + aa^*b)(a + ba^*b)^*$ since $R =\wedge R$
$= (\wedge + aa^*)b(a + ba^*b)$ since $RS + TS = (R + T)S$
$= (aa^*)b(a + ba^*b)^*$ since $\wedge + aa^* = aa^*$
$= a^*b(a + ba^*b)^*$.

Exercises on Identities on Regular Expression

1. $R(SR)^* = (RS)^*R$

   Let $x \in R(SR)^*$

   $\Rightarrow\ x = r[(s_1r_1)(s_2r_2)\cdots(s_kr_k)]$

   for $r_i \in R, s_i \in S, r \in R$

   $\Rightarrow\ x = (rs_1)(r_1s_2)\cdots(r_{k-1}s_k)r_k$

   $= [(rs_1)(r_1s_2)\cdots(r_{k-1}s_k)]r_k \in (RS)^*R$

   $\Rightarrow\ R(SR)^* \subset (RS)^*R$

   $y \in (RS)^*R$ … (1)

   $\Rightarrow = [(r_1s_1)(r_2s_2)\cdots(r_ks_k)]r$

   $= [r_1s_1r_2s_2\cdots r_ks_k]r$

   $= [r_1(s_1r_2)(s_2r_3)\cdots(s_{k-1}r_k)s_k]r$

   $= r_1(s_1r_2)(s_2r_3)\cdots(s_{k-1}r_k)(s_kr)$

   $= r_1[(s_1r_2)(s_2r_3)\cdots(s_{k-1}r_k)(s_kr)] \in R(SR)^*$

   $\Rightarrow\ (RS)^*R \subset R(SR)^*$ … (2)

   By (1) and (2)$R(SR)^* = (RS)^*R$

2. $(R + S)^* = (R^* + S^*)^*$

   Let $x \in (R^* + S^*)^*$

   $\Rightarrow\ x = [(r_1^* + s_1^*)(r_2^* + s_2^*)\cdots(r_k^* + s_k^*)]\ \forall r_i^* \in R^*, s_i^* \in S^*$.

   Now $r_i^* \in R^* \Rightarrow r_i^* = \left(r_{i_1}r_{i_2}\cdots r_{i_{k_i}}\right), i = 1, 2, \cdots, k$

   Similarly,

   $s_i^* \in S^* \Rightarrow s_i^* = \left(s_{i_1}s_{i_2}\cdots s_{i_{k_i}}\right), i = 1, 2, \cdots, k$

$$\therefore\ x = \left(r_{1_1}r_{1_2}\cdots r_{1_{k_1}} + s_{1_1}s_{1_2}\cdots s_{1_{k_1}}\right)$$
$$\left(r_{2_1}r_{2_2}\cdots r_{2_{k_1}} + s_{2_1}s_{2_2}\cdots s_{2_{k_2}}\right)$$
$$\left(r_{k_1}r_{k_2}\cdots r_{k_{k_k}} + s_{k_1}s_{k_2}\cdots s_{k_{k_k}}\right)$$

$r_{i_{k_i}} = r_{i_{k_i}} + \Lambda \in R + S$

$\Rightarrow r_{11}r_{12}\cdots r_{1k1} \in (R+S)^*$

$s_{i_{k_i}} = \Lambda + s_{i_{k_i}} \in R + S$

$\Rightarrow s_{11}s_{12}\cdots s_{1k1} \in (R+S)^*$

$\Rightarrow (R^* + S^*)^* \subset (R+S)^*$

Reversing the process we can prove,

$(R+S)^* \subset (R^* + S)^*$.

3. $(R^*S)^* = (R+S)^*S$

$x \in (R^*S^*)^*$

$\Leftrightarrow (r_{11}r_{12}\cdots r_{1k1}s_1)\,(r_{21}r_{22}\cdots r_{2k2}s_2)\cdots(r_{i1}r_{i2}\cdots r_{iki}s_i)$

$= (r_{11}r_{12}\cdots r_{1k1}s_1)\ \cdots (r_{i1}r_{i2}\cdots r_{iki})s_1$

$\Rightarrow x = (u_{11}u_{12}\cdots u_{1k1}u_1)\,(u_{21}u_{22}\cdots u_{2k2}u_2)\cdots(u_{i1}u_{i2}\cdots u_{iki})s_i$

[where $u_{11} = r_{11} + \Lambda \in R^* + S^*$, $u_{21} = r_{21} + \Lambda \in R^* + S^*, \cdots,$

$v_1 = \Lambda + s_1 \in R^* + S^*, v_2 = \Lambda + s_2 \in R^* + S^*$]

$\Leftrightarrow x \in (R^* + S^*)^*S$

$\Leftrightarrow x \in (R+S)^*S$ as $(R^* + S^*)^* = (R+S)^*$

Thus $(R^*S)^* = (R+S)^*S$

4. $(a+b+aa)^* = (a+b)^*$

$(a+b+aa)^* = \left((a+aa)+b\right)^*$ as $R+S = S+R$

$= [(a+aa)^* + b^*]^*$ as $(R+S)^* = (R^* + S^*)^*$

$= [a^* + b^*]^*$ as $(R+RR)^* = R^*$

$= [a+b]^*$

5. $a(a+b)^* + (a+b)^*aa(a+b)^* = [(b^*a)^*ab^*]^*$

$a(a+b)^* + (a+b)^*aa(a+b)^*$

$= [\Lambda + (a+b)^*a]a(a+b)^*]$

$= (a+b)^*aa(a+b)^*$

$= (b^*a)^*(ab^*)^*$ as $(R^*S)^* = (R+S)S$

$(RS^*)^* = R(R+S)^*$

6. $b^*(abb)^*[b^*(abb)^*]^* = (b + abb)^*b^*(abb)^*[b^*(abb)^*]^*$
$= (b^*(abb)^*)^*$ as $RR^* = R$
$= (b + abb)^*$ as $(R^*S^*)^* = (R + S)^*$

7. $(ba)^*b + (ba)^*(bb + a)[a + (ba)^*(bb + a)]^*b(ba)^*b$
$= [ba + (bba)(a^*b^*)]^*b(b^*a)b + (ba)^*(bb + a)$
$[(a + (ba)^*(bb + a)]^*b(ba)^*b$
$= [\wedge + (ba)^*(bb + a)([a + (ba)^*(bb + a)]^*b](ba)^*b$
$= (ba)^*(bb + a)([a + (ba)^*(bb + a)]^*b)(ba)^*b.$

Claim:

$[a + (ba)^*(bb + a)]^*b$
$= (a^*b)[(ba)^*(bb + a)a^*b]^*[a + (ba)^*(bb + a)]^*b$
$= a^*[(ba)^*(bb + a)a^*]^*b$ as $(R + S)^* = R^* = (SR^*)^*$
$= (a^*b)[(ba)^*(bb + a)a^*b]^*$ as $R^*S^*T = (R^*T)(ST)^*$

Then

LHS $= (ba)^*(bb + a)[(a^*b)(ba)^*(bb + a)a^*b]^*(ba)^*b$
$= (ba)(bb + a)(a^*b)(ba)^*(bb + a)a^*b^*(ba)^*b$
$= [(ba)^*(bb + a)a^*b]^*(ba)^*b$ as $RR^* = R$
$= [(ba)^*(bb + a)a^*b]b$[as $(R^*S^*)^*R^*$
$= (R + S)^*$ where $R = ba$
= RHS]

We close this chapter by stating (without proof) the following two theorems.

9.5.1 ***Theorem*** (T. Kleene, 1956). For any transition graph $T \in \Sigma, \exists$ a regular expression R such that $\tilde{R} = \tilde{T}$ and also for every regular expression R, $\exists$ a finite automation A such that $\tilde{A} = \tilde{R}$.

9.5.2 ***Theorem*** (Moore Equivalent Theorem) There exists an algorithm to determine whether two finite automata A and A' are equivalent or not.

CHAPTER 10

Theory of Fuzzy Sets

Advances in science and technology have made our modern society very complex and hence decision process has become increasingly vague and hard to analyze. Human brain has some special characteristic by which it is possible to learn and reason in a vague and fuzzy environment. It has the ability to arrive at decision based on imprecise, qualitative data in contrast to formal mathematics and formal logic which demands precise and quantitative data. Modern computers have over taken the brain's capacity to some extent, but still the human brain holds its charm by its ability to take decisions based on imprecise and qualitative data. Undoubtedly, in many areas of cognition human intelligence excels, the computer intelligence of today, and the development of tools capable of handling humanistic type of problem.

We have sufficient mathematical tools and computer based technology for analyzing and solving the problems embodied in deterministic and uncertain environment. Here uncertainty may arise from the probabilistic behavior of certain physical phenomena in mechanistic systems or due to qualitative data. We know the important role that vagueness and in exactitude play in human decision making, but until 1965 we did not know how the vagueness arising from subjectivity (which is inherent in human thought processes) can be modelled and analyzed.

John Von Neumann gave a serious thought in this direction during early fifties. John Von Neumann was born in Budapest, Hungary in 1903. From his childhood he had shown a remarkable ability and interest in scientific matters. In 1943 he became convinced and tried to convince others in many varied fields, that numerical calculations done on fast electronic computing devices, would substantially facilitate the solutions of many difficult, unsolved scientific problems. After the controversy with a group of selected engineers and mathematicians, Neumann built, at the Institute for advanced study, an experimental electronic calculator, popularly known as the JONIAC which eventually became the pilot model for similar machines all over the country. They tried to design the machine by initiating some of the known operations of the brain. For this he started reading neurology.

Early in 1955 he was invited by Yale University to give the Silliman lectures during the spring term of 1956. In Silliman lectures he wanted to give the improved idea of copying an extremely simplified model of the human brain for man-made machines. But in August 1955 he had developed severe pain in his left shoulder and it was diagnosed to be bone cancer. In spite of many efforts, he could not finish writing his planned lectures and died on February 8, 1957. But in order to pay respect to the departed soul, Yale University published his incomplete manuscript with a preface written by his wife. The last section of his unfinished manuscript was "The Language of Mathematics is not the Language of Human Beings".

Later in early sixties Professor L.A. Zadeh of the University of California, Berkeley thought in this direction. He asked "Can computer be made to think like human beings?"

In 1965, Prof. Zadeh laid the foundation of fuzzy set theory. Fuzzy set theory is the concept and technique which lay a form of mathematical precision to human thought processes that in many ways were imprecise and ambiguous by the standards of classical mathematics. The fuzzy theory has blossomed into a many faceted field of inquiry, drawing on and contributing to a wide spectrum of areas ranging from para mathematics to human perception and judgement. Fuzzy sets have been suggested for handling the imprecise real word problems by using truth values ranging between the usual "true" and "false".

There are classes of objects encountered in the real physical world which do not have precisely defined criteria of membership. For example, consider the collection of all small natural numbers or the class of all students with brilliant academic career or the collection of all tall men etc. All the classes defined above do not constitute sets in the usual mathematical sense. Yet, the fact remains that such imprecisely defined "classes" play an important role in human thinking particularly in the domains of pattern recognition, communication of information and abstraction.

L.A. Zadeh in 1965 explored a preliminary concept for these classes of ambiguities. He introduced the fuzzy set which is defined as a class with continnum of grades of membership. The behavior of fuzzy sets is similar in many senses with that of the ordinary sets but is more general than the later. It has much wider scope of applicability, particularly in the fields of pattern classification and information processing. It provides a natural way of dealing with problems in which the source of imprecision is the absence of sharply defined criteria of class of membership rather than the presence of random variables. Thus, a set can be called fuzzy set (but not vice-versa). Starting with the concepts in the set theory we will see how they lead to a generalized theory of fuzzy sets.

10.1 THE CONCEPT OF MEMBERSHIP

Let E be a set and $A \subseteq E$. Recall that 'an element x of E is a member of A' is indicated by using the symbol $\in$; that is $x \in A$ In order to indicate this membership one can use another concept also: the characteristic function μ_A from E taking values in $\{0.1\}$ is defined by,

$$\mu_A(x) = 1 \text{ if } x \in A$$
$$= 0 \text{ if } x \notin A$$

Let A and be B any two subsets of E.

We have

$$\mu_A(x) = 1 \text{ if } x \in A$$
$$= 0 \text{ if } x \notin A$$
$$\mu_B(x) = 1 \text{ if } x \in B$$
$$= 0 \text{ if } x \notin B$$
$$\mu_{A \cap B}(x) = 1 \text{ if } x \in A \cap B$$
$$= 0 \text{ if } x \notin A \cap B$$
$$\mu_{A \cup B}(x) = 1 \text{ if } x \in A \cup B$$
$$= 0 \text{ if } x \notin A \cup B$$

This allows us to write

$$\mu_{A \cap B} = \mu_A(x) \,.\, \mu_B(x)$$
$$\mu_{A \cup B} = \mu_A(x) + \mu_B(x)$$

where the operations $\cdot$ and $+$ correspond to tables in Figures 1 and 2 respectively and are called the Boolean product and Boolean sum.

·	0	1
0	0	0
1	0	1

Figure 1

+	0	1
0	0	1
1	1	1

Figure 2

Let A' be the complement of A with respect to E. Then,

$$\mu_A(x) = 1 \text{ if } x \in A$$
$$= 0 \text{ if } x \notin A$$
$$\mu_{A'}(x) = 1 \text{ if } x \in A'$$
$$= 0 \text{ if } x \notin A'$$

Further $A \cap A' = \emptyset, A \cup A' = E$. Thus we can write

$$\mu_{A'}(x) = 1 - \mu_A(x)$$

10.1.1 *Example*

1. Let $E = \{x_1, x_2, x_3, x_4, x_5\}$ and $A = \{x_1, x_2\}$
   We can represent the set A by accompanying the elements of E with their characteristic function values:
   $A = \{(x_1, 1), (x_2, 1), (x_3, 0), (x_4, 0), (x_5, 0)\}$
   Similarly, A' can be represented by,
   $A' = \{(x_1, 0), (x_2, 0), (x_3, 1), (x_4, 1), (x_5, 1)\}$
2. Let
   $B = \{(x_1, 1), (x_2, 0), (x_3, 1), (x_4, 0), (x_5, 1)\}$
   Then
   $A \cap B = \{(x_1, 1, 1), (x_2, 1, 0), (x_3, 0, 0), (x_4, 0, 0), (x_5, 0, 0)\}$
   $= \{(x_1, 0), (x_2, 0), (x_3, 0), (x_4, 0), (x_5, 0)\}$
   $A \cup B = \{(x_1, 1 + 0), (x_2, 0 + 0), (x_3, 0 + 0), (x_4, 0 + 0), (x_5, 0 + 0)\}$
   $= \{(x_1, 1), (x_2, 1), (x_3, 0), (x_4, 0), (x_5, 0)\}$

10.2 THE CONCEPT OF FUZZY SUB SET

Let E be a set. Let M be a totally ordered set (in particular M could be the interval [0, 1]). Then a fuzzy subset A of E is a set of ordered pairs,

$$\{(x, \mu_A(x))\}, \forall x \in E$$

where μ_A, is a function defined on E which takes its values in M. M is called a membership set and μ_A is called a membership or characteristic function. $\mu_A(x)$ is called the grade or degree or level of member-ship of x in A. If $M = \{0, 1\}$ then 'fuzzy subset' is an 'ordinary subset'.

The notion of fuzzy subsets is a generalization of the ordinary subsets to the case where the membership of elements are not clear cut and allows one to study, using mathematical structures, imprecise concept.

A set or a subset is denoted by $A, B, C, X, Y, Z, \ldots$. A fuzzy subset will be denoted by $\underset{\sim}{A}, \underset{\sim}{B}, \underset{\sim}{C}, \underset{\sim}{X}, \underset{\sim}{Y}, \underset{\sim}{Z}$ ...... If all the subsets under consideration are ordinary, then one may, suppress the small supplementary symbol ~. Membership and non-membership are denoted by $\in$ and $\notin$. Fuzzy membership and fuzzy non-membership will be denoted by $\underset{\sim}{\in}$ and $\underset{\sim}{\notin}$.

In certain cases where the membership set M is the closed interval $[0, 1]$ it may be convenient to represent the membership by accompanying the symbol $\in$ by a number from $[0, 1]$ placed beneath it. Thus

$x \underset{1}{\in} \underset{\sim}{A}$ means $x \in \underset{\sim}{A}$

$x \underset{0}{\in} \underset{\sim}{A}$ means $x \notin \underset{\sim}{A}$

$x \underset{0,8}{\in} \underset{\sim}{A}$ means $x \underset{\sim}{\in} \underset{\sim}{A}$ with degree 0.8

$x \underset{0,2}{\in} \underset{\sim}{A}$ means $x \underset{\sim}{\in} \underset{\sim}{A}$ with degree 0.2.

and so on.

Let $E = \{x_1, x_2, x_3, x_4, x_5\}$

and let $M = [0, 1]$.

Then $\underset{\sim}{A} = \{ (x_1 / 0.2), (x_2 / 0), (x_3 / 0.3), (x_4 / 1), (x_5 / 0.8)\}$

where x_i is an element of the reference set E and where the number placed after the bar is the value of the characteristic function for the element, is a fuzzy subset of E and is denoted by,

$\underset{\sim}{A} \subset E$

3. Let $E = \{a, b, c, d, e, f\}$ and $M = \{0, \frac{1}{2}, 1\}$.

   Then $\underset{\sim}{A} = \{(a/0), (b/1), (c/1/2), (d/0), (e/1/2), (f/0)\}$

   is a fuzzy subset of E.

4. Let N be the set of all natural numbers:
   $N = \{1, 2, 3, 4, \ldots \ldots\}$ and $M = [0, 1]$. Then
   $\underset{\sim}{A} = \{(0|1), (1|0,8), (2|0.6), (3|0.4), (4|0.2), (5|0), (6|0), \ldots \ldots \ldots\}$
   is a fuzzy subset of "small" natural numbers.
5. Let $E = \{0, 1, 2, 3, 4, 5, 6, 7, 8, 9\}$ and $M = [0, 1]$. Then
   $\underset{\sim}{A} = \{(0|0), (1|0.2), (2|0.3), (3|0), (4|1), (5|1), (6|0.8),$
   $(7|0.9), (8|0), (9|0)\}$
   is a fuzzy subset of E.

10.3 SIMPLE OPERATIONS ON FUZZY SUB SETS

Inclusion: Let E be a set and M its associated membership set. Let $\underset{\sim}{A}$ and $\underset{\sim}{B}$ be two fuzzy subsets of E. A is said to be included in B written as $\underset{\sim}{A} \subset \underset{\sim}{B}$ or $\underset{\sim}{A} \subseteq \underset{\sim}{B}$ if,

$$\mu_{\underset{\sim}{A}}(x) \leq \mu_{\underset{\sim}{B}}(x) \text{ for all } x \in E.$$

A is said to be strictly included in B written $\underset{\sim}{A} \subset\subset \underset{\sim}{B}$ or $\underset{\sim}{A} \subseteq\subset \underset{\sim}{B}$ if the above inequality is strict for at least one $x \in E$.

6. Let $E = \{x_1, x_2, x_3, x_4\}, M = [0, 1]$.

   $\underset{\sim}{A} = \{(x_1 | 0.4)(x_2 | 0.2), (x_3 | 0), (x_4 | 1)\}$

   $\underset{\sim}{B} = \{(x_1 | 0.3)(x_2 | 0), (x_3 | 0), (x_4 | 0)\}$

   One has $\underset{\sim}{B} \subset \underset{\sim}{A}$ since $0.3 < 0.4, 0 < 0.2, 0 = 0, 0 < 1$.

   One can write

   $$E = \{(x_1 | 1)(x_2 | 1), (x_3 | 1), (x_4 | 1)\}$$

   Thus, $E \subseteq E$ i.e., E is included in itself in the sense of the theory of fuzzy subsets.

Equality: Let E be a set, M its associated membership set, and $\underset{\sim}{A}$ and $\underset{\sim}{B}$ two fuzzy subsets of E. $\underset{\sim}{A}$ and $\underset{\sim}{B}$ are said to be equal if and only if,

$$\mu_{\underset{\sim}{A}}(x) = \mu_{\underset{\sim}{B}}(x) \text{ for all } x \in E.$$

This will be written as $\underset{\sim}{A} = \underset{\sim}{B}$. If for at least one $x \in E$, $\mu_{\underset{\sim}{A}}(x) \neq \mu_{\underset{\sim}{B}}(x)$ then $\underset{\sim}{A}$ and $\underset{\sim}{B}$ are said to be unequal and written as, $\underset{\sim}{A} \neq \underset{\sim}{B}$.

Complementation: Let E be a set and $M = [0, 1]$ its associated membership set. Let $\underset{\sim}{A}$ and $\underset{\sim}{B}$ be two fuzzy subsets of E. $\underset{\sim}{A}$ is the complement of $\underset{\sim}{B}$ or $\underset{\sim}{B}$ is the complement of $\underset{\sim}{A}$.

$$\mu_{\underset{\sim}{B}}(x) = 1 - \mu_{\underset{\sim}{A}}(x) \text{ for all } x \in E.$$

This will be denoted by $\underset{\sim}{B} = \underset{\sim}{A}'$ or $\underset{\sim}{A} = \underset{\sim}{B}'$.

One obviously has $(\underset{\sim}{A}')' = \underset{\sim}{A}$

7. Let $E = \{x_1, x_2, x_3, x_4, x_5, x_6\}, M = [0, 1]$.

   $\underset{\sim}{A} = \{(x_1 | 0.13), (x_2 | 0.61), (x_3 | 0), (x_4 | 0), (x_5 | 1), (x_6 | 0.03)\}$

   $\underset{\sim}{B} = \{(x_1 | 0.87), (x_2 | 0.39), (x_3 | 1), (x_4 | 1), (x_5 | 0), (x_6 | 0.97)\}$

   Then $\underset{\sim}{A}' = \underset{\sim}{B}$

Intersection: Let E be a set and $M = [0, 1]$ its associated membership set and let $\underset{\sim}{A}$ and $\underset{\sim}{B}$ be two fuzzy subsets of E. Then the intersection of $\underset{\sim}{A}$ and $\underset{\sim}{B}$ denoted by $\underset{\sim}{A} \cap \underset{\sim}{B}$ is defined as the largest fuzzy subset contained at the same time in $\underset{\sim}{A}$ and $\underset{\sim}{B}$. That is

$$\mu_{\underset{\sim}{A} \cap \underset{\sim}{B}}(x) = \min\{\mu_{\underset{\sim}{A}}(x), \mu_{\underset{\sim}{B}}(x) \text{ for all } x \in E.$$

8. Let $\underset{\sim}{A} = \{(x_1| \, 0.2), (x_2| \, 0.7), (x_3| \, 1), (x_4 \, |0), (x_5| \, 0.5)\}$

   $\underset{\sim}{B} = \{(x_1| \, 0.5), (x_2| \, 0.3), (x_3| \, 1), (x_4 \, |0.1), (x_5| \, 0.5)\}$

   $\underset{\sim}{A} \cap \underset{\sim}{B} = \{(x_1| \, 0.2), (x_2| \, 0.3), (x_3| \, 1), (x_4 \, |0), (x_5| \, 0.5)\}$

Union: Let E be a set and $M = [0, 1]$ its associated membership set and let $\underset{\sim}{A}$ and $\underset{\sim}{B}$ be two fuzzy subsets of E. The union of $\underset{\sim}{A}$ and $\underset{\sim}{B}$ denoted by $A \cup B$ is defined as the smallest fuzzy subset that contains both $\underset{\sim}{A}$ and $\underset{\sim}{B}$. That is,

$$\mu_{\underset{\sim}{A} \cup \underset{\sim}{B}}(x) = \max\{\mu_{\underset{\sim}{A}}(x), \mu_{\underset{\sim}{B}}(x)\} \text{ for all } x \in E.$$

In Example 8 we have

Let $\underset{\sim}{A} \cup \underset{\sim}{B} = \{(x_1| \, 0.5), (x_2| \, 0.7), (x_3| \, 1), (x_4 \, |0.1), (x_5| \, 0.5)\}$

Difference: Let E be a set and $M = [0, 1]$ its associated membership set and let $\underset{\sim}{A}$ and $\underset{\sim}{B}$ be two fuzzy subsets of E. The difference of $\underset{\sim}{A}$ and $\underset{\sim}{B}$ is defined by the relation,

$$\underset{\sim}{A} - \underset{\sim}{B} = \underset{\sim}{A} \cap \underset{\sim}{B}'$$

Considering Example 8 we have,

8. $\underset{\sim}{A} - \underset{\sim}{B}' = \{(x_1| \, 0.2), (x_2| \, 0.7), (x_3| \, 0), (x_4 \, |0), (x_5| \, 0.5)\}$

   Note that, except in particular cases,

   $\underset{\sim}{A} - \underset{\sim}{B} \neq \underset{\sim}{B} - \underset{\sim}{A}$

10.4 VISUAL REPRESENTATION OF FUZZY SETS

We know that ordinary subsets can be represented by Venn diagram. For fuzzy subsets one can also construct a visual representation.

Consider a rectangle (Figure 1) with the values $\mu_{\underset{\sim}{A}}(x)$ as ordinate and as abscissa the elements of E in an arbitrary order (if there is some total order in E that order has to be taken). In Figure 1 the membership of each element is represented by its ordinate. The shaded part represents the fuzzy subsets

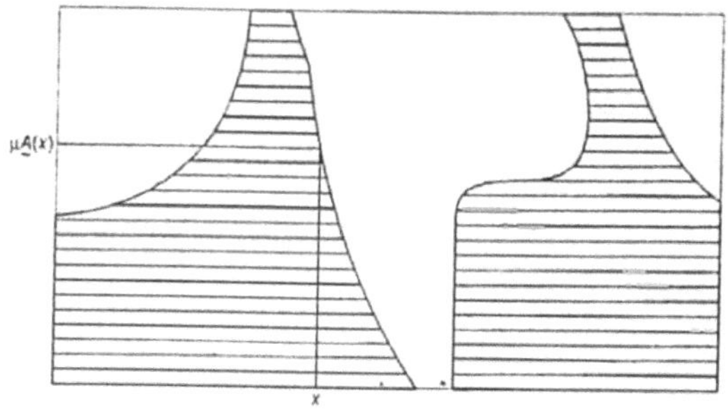

Figure 1

$\underset{\sim}{A} \subset E$. With this representation one can visualize the various simple operations such as: Complementation, union, intersection etc. for fuzzy subsets.

10.5 FUZZY SUBSET FOR E AND M FINITE

Let E and M be finite. We first recall the definition of the set of all subsets (or power set) of a set by considering a simple example. Let $E = \{x_1, x_2, x_3\}$.

Then the power set of $E = \varphi(E)$, is represented as follows:

$$\varphi(E) = \{\emptyset, \{x_1\}, \{x_2\}, \{x_3\}, \{x_1, x_2\}, \{x_1, x_3\}\ \{(x_2, x_3), E\}$$

There are $2^3 = 8$ elements in this set. More generally if,

$$E = \{x_1, x_2, x_3, \ldots\ldots\ldots.., x_n\}$$

Has n elements, $\varphi(E)$ will have 2^nelements.

For a fuzzy subset, the power set or "the set of fuzzy subsets" is presented in slightly different manner. Let

Let $E = \{x_1, x_2\}$ and $M = [0, 1/2, 1]$.

Then the set of fuzzy subsets, $\varphi(\underset{\sim}{E})$ will be

$$\varphi(\underset{\sim}{E}) = \{(x_1|\ 0), (x_2|\ 0)\}, \{(x_1|\ 0), (x_2\ |0.5)\}$$

$$\{(x_1|\ 0.5), (x_2|\ 0)\}, \{(x_1|\ 0.5), (x_2\ |0.5)\}$$

$$\{(x_1|\ 0), (x_2|\ 1)\}, \{(x_1|\ 1), (x_2\ |0)\}$$

$$\{(x_1|\ 1), (x_2|\ 0.5)\}, \{(x_1|\ 0.5), (x_2\ |1)\}$$

$$\{(x_1(|\ 1), (x_2|\ 1)\}$$

More generally, if

$$Card\ E = n \text{ and } Card\ M = m$$

where card means "cardinality of", that is, the number of elements for a finite set, then

$$\text{card }(\varphi(\underset{\sim}{E})) = m^n$$

The set $\varphi(\underset{\sim}{E})$ contains 2^n ordinary subsets. Note that one always has $\varphi(E) \subset \varphi(\underset{\sim}{E})$.

10.6 PROPERTIES OF FUZZY SUBSET

Recall that the power set E of an ordinary set has the following properties: for $A \subset E, B \subset E, C \subset E$ and we have,

1. $A \cap B = B \cap A$,
2. $A \cup B = B \cup A$,
3. $(A \cap B) \cap C = A \cap (B \cap C)$,
4. $(A \cup B) \cup C = A \cup (B \cup C)$,
5. $A \cap A = A$,
6. $A \cup A = A$,
7. $A \cap (B \cup C) = (A \cap B) \cup (A \cap C)$,
8. $A \cup (B \cap C) = (A \cup B) \cap (A \cup C)$,
9. $A \cap A' = \emptyset$,
10. $A \cup A' = E$,
11. $A \cap \varphi = \varphi$
12. $A \cup \emptyset = A$,
13. $A \cap E = A$,
14. $A \cup E = E$,
15. $(A')' = A$,
16. $(A \cap B)' = A' \cup B'$,
17. $(A \cup B)' = A' \cap B'$

    If $\underset{\sim}{A}, \underset{\sim}{B}, \underset{\sim}{C}$ are fuzzy subsets of E all the properties 1–17 hold except 9 and 10, these two hold for ordinary subsets only. For fuzzy subsets we have.

18. $\underset{\sim}{A} \cap \underset{\sim}{B} = \underset{\sim}{B} \cap \underset{\sim}{A}$
19. $\underset{\sim}{A} \cup \underset{\sim}{B} = \underset{\sim}{B} \cup \underset{\sim}{A}$
20. $(\underset{\sim}{A} \cap \underset{\sim}{B}) \cap \underset{\sim}{C} = \underset{\sim}{A} \cap (\underset{\sim}{B} \cap \underset{\sim}{C})$

21. $(\underset{\sim}{A} \cup \underset{\sim}{B}) \cup \underset{\sim}{C} = \underset{\sim}{A} \cup (\underset{\sim}{B} \cup \underset{\sim}{C})$
22. $\underset{\sim}{A} \cap \underset{\sim}{A} = \underset{\sim}{A}$
23. $\underset{\sim}{A} \cup \underset{\sim}{A} = \underset{\sim}{A}$
24. $\underset{\sim}{A} \cap (\underset{\sim}{B} \cup \underset{\sim}{C}) = (\underset{\sim}{A} \cap \underset{\sim}{B}) \cup (\underset{\sim}{A} \cap \underset{\sim}{C})$
25. $\underset{\sim}{A} \cup (\underset{\sim}{B} \cap \underset{\sim}{C}) = (\underset{\sim}{A} \cup \underset{\sim}{B}) \cap (\underset{\sim}{A} \cup \underset{\sim}{C})$
26. $\underset{\sim}{A} \cap \emptyset = \emptyset$ where $\emptyset$ is the ordinary set such that $\mu_{\phi}(x_i) = 0$ for all $x_i \in E$.
27. $\underset{\sim}{A} \cup \emptyset = \underset{\sim}{A}$
28. $\underset{\sim}{A} \cap \mathrm{E}\ \underset{\sim}{A}$, where E is such that $\mu_E(x_i) = 1$ for all $x_i \in E$. (i.e. the (ordinary) universal reference set).
29. $\underset{\sim}{A} \cup E = E$
30. $(\underset{\sim}{A}')' = \underset{\sim}{A}$
31. $(\underset{\sim}{A} \cap \underset{\sim}{B})' = \underset{\sim}{A}' \cup \underset{\sim}{B}'$
32. $(\underset{\sim}{A} \cup \underset{\sim}{B})' = \underset{\sim}{A}' \cap \underset{\sim}{B}'$

10.7 ALGEBRAIC PRODUCT AND SUM OF TWO FUZZY SUBSETS

Let E be a set and $M = [0, 1]$ its associated membership set. Let $\underset{\sim}{A}$ and $\underset{\sim}{B}$ be two fuzzy subsets of E. We define the algebraic product of $\underset{\sim}{A}$ and $\underset{\sim}{B}$, denoted by $\underset{\sim}{A} \cdot \underset{\sim}{B}$ and algebraic sum of $\underset{\sim}{A}$ and $\underset{\sim}{B}$ denoted by $\underset{\sim}{A} \hat{+} \underset{\sim}{B}$ in the following ways:

$$\mu_{\underset{\sim}{A} \cdot \underset{\sim}{B}}(x) = \mu_{\underset{\sim}{A}}(x) \cdot \mu_{\underset{\sim}{B}}(x), \ \forall x \in E$$

$$\mu_{\underset{\sim}{A} \hat{+} \underset{\sim}{B}}(x) = \mu_{\underset{\sim}{A}}(x) + \mu_{\underset{\sim}{B}}(x) - \mu_{\underset{\sim}{A}}(x) \cdot \mu_{\underset{\sim}{B}}(x), \ \forall x \in E$$

9. Let

$$\underset{\sim}{A} = \{(x_1 | \, 0.2), (x_2 | \, 0.7), (x_3 | \, 1), (x_4 \, | 0), (x_5 | \, 0.5)\}$$

$$\underset{\sim}{B} = \{(x_1 | \, 0.5), (x_2 | \, 0.3), (x_3 | \, 1), (x_4 \, | 0.1), (x_5 | \, 0.5)\}$$

Then

$$\underset{\sim}{A} \cdot \underset{\sim}{B} = \{(x_1 | \, 0.10), (x_2 | \, 0.21), (x_3 | \, 1), (x_4 \, | 0), (x_5 | \, 0.25)\}$$

$$\underset{\sim}{A} \hat{+} \underset{\sim}{B} = \{(x_1 | \, 0.60), (x_2 | \, 0.79), (x_3 | \, 1), (x_4 \, | 0.1), (x_5 | \, 0.75)\}$$

Note that if $M = \{0, 1\}$ i.e. if we are discussing about ordinary subsets, then

$$A \cap B = A \cdot B, A \cup B = A \overset{\wedge}{+} B$$

One can verify the following properties on the power set of fuzzy subsets with the two operations $\cdot$ $\overset{\wedge}{+}$ and. It may be noted here properties similar to 5 and 6, 7 and 8 (distributivity), 9 and 10 with $\cup$ and $\cap$ replaced by $\overset{\wedge}{+}$ and, ordinary subsets replaced by fuzzy subsets are not true. What are true are only the following:

33. $\underset{\sim}{A} \cdot \underset{\sim}{B} = \underset{\sim}{B} \cdot \underset{\sim}{A}$

34. $\underset{\sim}{A} \overset{\wedge}{+} \underset{\sim}{B} = \underset{\sim}{B} \overset{\wedge}{+} \underset{\sim}{A}$

35. $(\underset{\sim}{A} \cdot \underset{\sim}{B}) \cdot \underset{\sim}{C} = \underset{\sim}{A} \cdot (\underset{\sim}{B} \cdot \underset{\sim}{C})$

36. $(\underset{\sim}{A} \overset{\wedge}{+} \underset{\sim}{B}) \overset{\wedge}{+} \underset{\sim}{C} = \underset{\sim}{A} \overset{\wedge}{+} (\underset{\sim}{B} \overset{\wedge}{+} \underset{\sim}{C})$

37. $\underset{\sim}{A} \cdot \emptyset = \emptyset$

38. $\underset{\sim}{A} \overset{\wedge}{+} \varnothing = \underset{\sim}{A}$

39. $\underset{\sim}{A} \cdot E = \underset{\sim}{A}$

40. $\underset{\sim}{A} \overset{\wedge}{+} E = E$

41. $(\underset{\sim}{A} \cdot \underset{\sim}{B})' = (\underset{\sim}{A})' \overset{\wedge}{+} (\underset{\sim}{B})'$

42. $(\underset{\sim}{A} \overset{\wedge}{+} \underset{\sim}{B})' = (\underset{\sim}{A})' \cdot (\underset{\sim}{B})'$

EXERCISES

1. Find $\underset{\sim}{A} \cup \underset{\sim}{B}$, $\underset{\sim}{A} \cap \underset{\sim}{B}$, $\underset{\sim}{A} - \underset{\sim}{B}$, $\underset{\sim}{B} - \underset{\sim}{A}$ in the following cases:
   (i) $\underset{\sim}{A} = \{(x_1| 0.2), (x_2| 0.8), (x_3| 0.6)\}$
   $\underset{\sim}{B} = \{(x_1| 0.4), (x_2| 0.6), (x_3| 1.0)\}$.
   (ii) $\underset{\sim}{A} = \{(x_1| 1.0), (x_2| 0.0), (x_3| 0.5), (x_4| 0.7)\}$
   $\underset{\sim}{B} = \{(x_1| 0.8), (x_2| 1.0), (x_3| 0.0), (x_4| 0.6)\}$
2. If $E = \{x_1, x_2, x_3, x_4\}$ and $M = [0, 1]$. Find $\underset{\sim}{A}'$, $\underset{\sim}{B}'$ in 1(i) and (ii).
3. Prove De Morgan's Laws for fuzzy subsets.

4. Find $\underset{\sim}{A} \cdot \underset{\sim}{B}$ and $\underset{\sim}{A} \mathbin{\overset{\wedge}{+}} \underset{\sim}{B}$ in 1(i) and (ii).

5. Find the fuzzy power set of the following:
   $S = \{x_1, x_2\}, M = \{0, 0.6, 1\}$,
   $S = \{x_1, x_2, x_3\}, M = \{0.5, 0.6\}$

Bibliography

1. Bernays, J., Symbolic Logic, 2 (1973) 65–77.
2. Birkhoff, G. and Thomas C. Bartee, Modern Applied Algebra, Mc Graw Hill International C.B.S Publishers and Distributors, Delhi.
3. Cantor, G., Math. Annalen, 20 (1882), 113–121.
4. Fraleigh, J.B., A First Course in Abstract Algebra, Reading, Addition – Wesley (1967) Fourth Print 1972.
5. Grill, A., Applied Algebra for Computer Science.
6. Joseph A. Gallian, Contemporary Abstract Algebra, Narosa Publishing House, New Delhi 1999.
7. Herstein, I.N., Topics in Algebra, John-Wiley and Sons, Wiley Eastern Ltd., Narosa/Vikas Publishing Co.
8. Kolman, Bernard and Robert, C. Busby, Discrete Mathematical Structures for Computer Science, Prentice – Hall of India 1988.
9. Kaufman, A., Introduction to the Theory of Fuzzy Subsets, Vol. I, Academic Press, New York.
10. Kreyszig, E., Advanced Engineering Mathematics.
11. Levy, L.S., Discrete Structures for Computer Science, John Wiley.
12. Liu C.L., Elements of Discrete Mathematics, McGraw-Hill Book Company.
13. Lipschitz, S., Discrete Mathematics, Schaum's Outline Series.
14. Kenneth, H. Rosen, Discrete Mathematics and its Applications, Tata Mcgraw Hill Publishing Co. Ltd., New Delhi 2003.
15. Russell and Whitehead, Principia Mathematica (1910).
16. Stanat, D.F. and McAlister, D.F., Discrete Mathematics in Computer Science, Prentice-Hall Englewood, Cliffs, NJ (1977).
17. Simmons, G.F., Introduction to Topology and Modern Analysis, McGraw-Hill, New York (1963).
18. Tremablay, J.P. and R. Manoher, Discrete Mathematical Structures with Applications to Computer Science, McGraw-Hill, NY (1977).
19. Zadeh, L.A., Fuzzy Sets, Information and Control, 8 (1965), 338–353.

20 Zermelo, Mathematische Annalen, 65 (1908), 261–281.

Index